Global Warming Cycles

GLOBAL
WARMING

Global Warming Cycles

Ice Ages and Glacial Retreat

Julie Kerr Casper, Ph.D.

✔ Facts On File
An imprint of Infobase Publishing

GLOBAL WARMING CYCLES: Ice Ages and Glacial Retreat

Facts On File, Inc.
An imprint of Infobase Publishing
132 West 31st Street
New York NY 10001

Library of Congress Cataloging-in-Publication Data

Casper, Julie Kerr.
    Global warming cycles : ice ages and glacial retreat / Julie Kerr Casper.
        p. cm.—(Global warming)
    Includes bibliographical references and index.
    ISBN 978-0-8160-7262-0
    1. Global warming. I. Title.
    QC981.8.G56C37 2010
    551.6—dc22   2008053347

Facts On File books are available at special discounts when purchased in bulk quantities for businesses, associations, institutions, or sales promotions. Please call our Special Sales Department in New York at (212) 967-8800 or (800) 322-8755.

You can find Facts On File on the World Wide Web at http://www.factsonfile.com

Text design by Erik Lindstrom
Illustrations by Melissa Ericksen and Sholto Ainslie
Photo research by the author

Printed in the United States of America

Bang Hermitage 10 9 8 7 6 5 4 3 2 1

This book is printed on acid-free paper.

CONTENTS

PREFACE

We do not inherit the Earth from our ancestors—
we borrow it from our children.

This ancient Native American proverb and what it implies resonates today as it has become increasingly obvious that people's actions and interactions with the environment affect not only living conditions now, but also those of many generations to follow. Humans must address the effect they have on the Earth's climate and how their choices today will have an impact on future generations.

Many years ago, Mark Twain joked that "Everyone talks about the weather, but no one does anything about it." That is not true anymore. Humans are changing the world's climate and with it the local, regional, and global weather. Scientists tell us that "climate is what we expect, and weather is what we get." Climate change occurs when that average weather shifts over the long term in a specific location, a region, or the entire planet.

Global warming and climate change are urgent topics. They are discussed on the news, in conversations, and are even the subjects of horror movies. How much is fact? What does global warming mean to individuals? What should it mean?

The readers of this multivolume set—most of whom are today's middle and high school students—will be tomorrow's leaders and scientists. Global warming and its threats are real. As scientists unlock the mysteries of the past and analyze today's activities, they warn that future generations may be in jeopardy. There is now overwhelming evidence that human activities are changing the world's climate. For thousands of years, the Earth's atmosphere has changed very little; but today, there are problems in keeping the balance. Greenhouse gases are being added to the atmosphere at an alarming rate. Since the Industrial Revolution (late 18th, early 19th centuries), human activities from transportation, agriculture, fossil fuels, waste disposal and treatment, deforestation, power stations, land use, biomass burning, and industrial processes, among other things, have added to the concentrations of greenhouse gases.

These activities are changing the atmosphere more rapidly than humans have ever experienced before. Some people think that warming the Earth's atmosphere by a few degrees is harmless and could have no effect on them; but global warming is more than just a warming—or cooling—trend. Global warming could have far-reaching and unpredictable environmental, social, and economic consequences. The following demonstrates what a few degrees' change in the temperature can do.

The Earth experienced an ice age 13,000 years ago. Global temperatures then warmed up 8.3°F (5°C) and melted the vast ice sheets that covered much of the North American continent. Scientists today predict that average temperatures could rise 11.7°F (7°C) during this century alone. What will happen to the remaining glaciers and ice caps?

If the temperatures rise as leading scientists have predicted, less freshwater will be available—and already one-third of the world's population (about 2 billion people) suffer from a shortage of water. Lack of water will keep farmers from growing food. It will also permanently destroy sensitive fish and wildlife habitat. As the ocean levels rise, coastal lands and islands will be flooded and destroyed. Heat waves could kill tens of thousands of people. With warmer temperatures, outbreaks of diseases will spread and intensify. Plant pollen mold spores in the air will increase, affecting those with allergies. An increase in severe weather could result in hurricanes similar or even stronger than Katrina in 2005, which destroyed large areas of the southeastern United States.

Higher temperatures will cause other areas to dry out and become tinder for larger and more devastating wildfires that threaten forests, wildlife, and homes. If drought destroys the rain forests, the Earth's delicate oxygen and carbon balances will be harmed, affecting the water, air, vegetation, and all life.

Although the United States has been one of the largest contributors to global warming, it ranks far below countries and regions—such as Canada, Australia, and western Europe—in taking steps to fix the damage that has been done. Global Warming is a multivolume set that explores the concept that each person is a member of a global family who shares responsibility for fixing this problem. In fact, the only way to fix it is to work together toward a common goal. This seven-volume set covers all of the important climatic issues that need to be addressed in order to understand the problem, allowing the reader to build a solid foundation of knowledge and to use the information to help solve the critical issues in effective ways. The set includes the following volumes:

Climate Systems
Global Warming Trends
Global Warming Cycles
Changing Ecosystems
Greenhouse Gases
Fossil Fuels and Pollution
Climate Management

These volumes explore a multitude of topics—how climates change, learning from past ice ages, natural factors that trigger global warming on Earth, whether the Earth can expect another ice age in the future, how the Earth's climate is changing now, emergency preparedness in severe weather, projections for the future, and why climate affects everything people do from growing food, to heating homes, to using the Earth's natural resources, to new scientific discoveries. They look at the impact that rising sea levels will have on islands and other areas worldwide, how individual ecosystems will be affected, what humans will lose if rain forests are destroyed, how industrialization and pollution puts peoples' lives at risk, and the benefits of developing environmentally friendly energy resources.

The set also examines the exciting technology of computer modeling and how it has unlocked mysteries about past climate change and global warming and how it can predict the local, regional, and global climates of the future—the very things leaders of tomorrow need to know *today.*

We will know only what we are taught;
We will be taught only what others deem is important to know;
And we will learn to value that which is important.
— Native American proverb

ACKNOWLEDGMENTS

Global warming may be one of the most important issues influencing your decisions in your lifetime. The decisions you make on energy sources and daily conservation practices will determine not only the quality of your life but also those of your future descendants.

I cannot stress enough how important it is to gain a good understanding of global warming: what it is, why it is happening, how it can be slowed down, why everybody is contributing to the problem, and why *everybody* needs to be an active part of the solution.

I would sincerely like to thank several of the federal government agencies that research, educate, and actively take part in dealing with the global warming issue, in particular, the National Aeronautics and Space Administration (NASA), the National Oceanic and Atmospheric Administration (NOAA), the Environmental Protection Agency (EPA), and the U.S. Geological Survey (USGS) for providing an abundance of resources and outreach programs on this important subject. I would especially like to acknowledge the years of leadership and research provided by Dr. James E. Hansen of NASA's Goddard Institute for Space Studies (GISS). His pioneering efforts over the past 20 years have enabled other scientists, researchers, and political leaders worldwide to better understand the scope of the scientific issues involved at a critical point in time when action must be taken before it is too late.

I also give special thanks to former congressman and vice president Al Gore and current California governor Arnold Schwarzenegger for their diligent efforts toward bringing the global warming issue to the public's attention. I would also like to acknowledge and give thanks to the many wonderful universities in the United States, Great Britain, Canada, and Australia, as well as private organizations, such as the World Wildlife Fund, that diligently strive to educate others and help toward finding a solution to this very real problem.

I want to give a huge thanks to my agent, Jodie Rhodes, for her assistance, guidance, and efforts; and also to Frank K. Darmstadt, my editor, for all his hard work, dedication, support, helpful advice, and attention to detail. His efforts in bringing this project to life were invaluable. Thanks also to the copyediting and production departments for their assistance and the outstanding quality of their work.

INTRODUCTION

Earth has always exhibited patterns of heating up and cooling down. At some points in time, many areas of Earth are shrouded in blankets of ice, with ice caps and glaciers dominating the landscape. In fact, certain areas on Earth have been covered multiple times in the past with prominent glaciers for millions of years. Then, during the intervals between these ice ages, Earth's temperatures have warmed up, and the ice has melted and receded. Since the last ice age, which ended just over 10,000 years ago (a short time, geologically speaking), Earth's climate has been relatively stable, with just a few fluctuations—at least until the beginning of the industrial revolution in the 1700s, when the climate began to increase in temperature.

Earth's natural climate is always in flux, adapting and adjusting to internal and external processes. Just as natural processes can have an impact, so can the behavior of humans. Added to this are the external factors relating to variations in solar radiation and cyclic changes in Earth's orbit. All of these inputs—these "forcings"—cause the climate to react in a certain way. The time element is also variable: Some climate changes take thousands of years to take effect; others, only decades. All of the contributing factors of climate, such as temperature, precipitation, atmospheric circulation, ocean circulation, and the distribution of landforms are pieces in the master climatic puzzle, helping shape it over eons of geologic time.

Some of the most commonly studied elements of climate are the ice ages, because ice sheets and glaciers respond so rapidly to the forcings of climate change. Earth's climate changes whenever the amount of energy stored by the climate system varies. This usually happens when there is a change in Earth's global energy balance—the incoming energy from the Sun being balanced with outgoing heat from Earth. There are some natural conditions that can upset this balance, such as changes in Earth's orbit, changes in the composition of the atmosphere, or changes in ocean currents. Today, however, most of the major disruptions are due to human-caused pollution from the emission of greenhouse gases. When any of these mechanisms are significant enough to change the energy balance, they are said to "force" the climate to change. Climate scientists have adopted the terms *climate forcing* and *forcings* to explain these mechanisms' influence on the climate.

Ice ages have also occurred throughout Earth's history in a cyclic pattern and left landforms and other evidence behind, enabling scientists to study them and obtain a detailed, reliable glimpse into the past. By analyzing the repeated advance and retreat of glaciers and the effects and contribution of the environment around it, scientists gain a better understanding of how, when, and why ice ages have occurred. In fact, much climatic research occurring today is related to the glacial and interglacial cycles of the current ice age. Because climate acts as a huge working system, scientists can analyze past evidence and understand the processes that occurred long ago by comparing them to processes that occur today.

As a result of the continuing warmth worldwide, Earth is feeling the effects, though not all results are the same. Some areas are becoming warmer and wetter; some, warmer and drier. Some are experiencing flooding; others, drought. Ocean circulation patterns are changing, wind patterns are altering, and cloud cover is affecting how much solar energy reaches Earth's surface and contributes to the greenhouse effect. Sea ice and glaciers are thinning, and temperatures are expected to keep rising.

Even though the temperature's changing a few degrees may not seem a big concern, it is. During the last ice age, for instance, global temperatures were only about 6.7 to 10° Fahrenheit (4–6° Celsius) cooler than they are today. To get a handle on the climate changes today, it is vitally important that climatologists gain a good understanding of the mechanisms that caused past climate change, putting the Earth repeatedly into and out of ice ages. By understanding these natural cycles, they are better able to compare them to the differences of today's rapid global warming, caused principally by the behavior of humans. This knowledge, in turn, allows them to understand the science of global warming better, a necessary step to solving the problem.

Global Warming Cycles: Ice Ages and Glacial Retreat, one volume in the Global Warming set, focuses on these issues. Chapter 1 looks at what ice ages are and why they occur. Each glaciated period on Earth varies in its intensity, and none is like another. Therefore, if the issue of global warming is such a key topic today, why are some scientists worried about the Earth entering another ice age? My goal is to show you the relationship between the two issues and how a natural cycle can indeed be affected by a human-influenced process. This chapter also presents the many physical factors that shape the Earth's climate and which ones are out of our control and which ones humans have a direct impact on. Finally, we take a look at a well-known, extremely controversial theory called the "hockey stick theory" and present the reasons for the controversy, with the intent that you, as the reader, can decide for yourself what it means.

Chapter 2 focuses on glacial retreat. It looks at the evidence that past glaciers have left, which can clue scientists in today as to what Earth's climate was once like. It looks at how scientists use landform clues from the past to assess global warming and then apply that knowledge to the future in order to help prepare current and future society for what is to come. If scientists do not understand these correlations, there is no way that land managers and politicians will be able to plan ahead and provide for the safety and well-being of towns and cities.

Chapter 3 discusses ice sheets and the phenomenon of isostasy. It looks at research done in Antarctica and Greenland and the consequences of recent breakup of large pieces of the world's largest ice sheets. Both isostasy and the melting of ice sheets affect sea level. This is especially important for every inhabitant worldwide who lives near the coast. If sea levels rise, many populated areas will become submerged and destroyed; therefore, it is important to understand that the consequences of global warming can potentially do extreme damage to present-day developments.

The following two chapters look at the world's oceans as a massive heat transport system. They illustrate how oceans are able to retain so much heat; the role they have played over the years to cancel out the effects of global warming; how they have made some areas on Earth habitable, where otherwise they may not be; and why scientists need to diligently monitor them so that balances do not get easily tipped. I also introduce computer-modeling techniques and other technology available today to help climatologists better predict the conditions of coastlines.

Chapter 6 focuses on the science of abrupt climate change, an issue that has scientists worldwide very concerned with the future of the Earth and global warming. When the film *The Day After Tomorrow* was released, the thought of, and the potential impacts of, abrupt climate change frightened many people. This chapter looks at what types of things can really happen and, given the condition of Earth today, what the chances are of us experiencing a climate change of significant consequence.

A discussion on tropical cyclones and other severe weather follows, in chapter 7. The rates of incidence will be a consequence of further global warming. Also within the chapter are some fascinating statistics about which areas of the country have the costliest hurricanes and where in the United States you do not want to be if a hurricane makes landfall. The final two chapters, meanwhile focus on what various climate experts have to say on the subject, conclusions, and a glance into the future. Each of the climate research scenarios represented touches on a current aspect of the ocean and global warming being done by leading scientists around the world. Not all scientists agree; disagreement is common in the pursuit of understanding. It is as valuable to know what issues are right as it is to know which ones are wrong; that is how the knowledge base of a scientific field grows. While we learn a great deal from scientists, scientists also learn a great deal from one another. This data sharing is an extremely important role that each good scientist needs to play in order to progress.

Global warming is a huge topic, involving many facets of society: climatology, hydrology, geography, geology, biology, botany, ecology, environmentalism, physics, chemistry, economics, and political science, to name just a few. My goal with regard to this set is to introduce you, the reader, to the entire spectrum of issues so that responsible, knowledgeable decisions can be made in the future. A deep knowledge of global warming can only come one step at a time. As you read, you will make progress toward unlocking the door to understanding the issues, and as the author, I would like to give you the key.

Ice Ages

Scientists have determined through proxy evidence that Earth has experienced several natural ice age intervals throughout its history. Understanding what they are and why they occur helps them better understand the forces at work today shaping the *climate*. It also gives them insights into what humans can expect in the future. This chapter first studies ice ages, why they occur, and how scientists are able to uncover this information about events that happened millions of years ago. It also looks at the concept of climate cycles and prominent cooling events, presents the hockey stick theory and what that means for future climate, and finally, addresses the concept of global dimming and whether a future ice age may be possible.

ICE AGES—WHAT THEY ARE AND WHY THEY OCCUR

Ice ages occur when temperatures remain cool for extended lengths of time. When long intervals of cold prevail, the ice does not melt; instead, ice caps and glaciers are able to advance, extending into lower latitudes, farther from the polar regions. Scientists have discovered evidence for this worldwide. For instance, during the last ice age 20,000 years ago, ice sheets advanced from the North Pole south to cover Canada, the New England states, the upper portions of the Midwest, Alaska, Greenland, Iceland, Scandinavia, most of Great Britain and Ireland, and the northwestern portion of Russia. Evidence confirming these events in the form of glacial landforms and deposits still exists today and is studied by climatologists worldwide.

When climatologists look backward in time and piece together evidence of Earth's past climate, they are able to identify the cyclic nature of glacial periods. Currently, they have determined that over the past millions of years, there have been multiple ice age cycles. Each glacial cycle is separated by a warmer period—referred to as an "interglacial period." During the Phanerozoic (the past 545 million years), Earth has passed through eight great climate cycles, each lasting 50 to 90 million years. There were also ice ages 52, 36, 20, 15, 7, 5, and 3 million years ago. During the past million years alone, there were eight to 10 ice ages, each only about 100,000 years long, interspersed with short, warm interglacial periods about 10,000 years long. Earth is currently experiencing an interglacial climate.

Each glacial period varies in intensity; they are not all alike. Ice ages can last thousands to millions of years and involve major ice sheets that cover continents and also include short intervals when glaciers reach their maximum extents, called "glaciations."

Scientists have determined that over Earth's past billion years, there have been several glacial advances and retreats. They are grouped into the following four distinct time intervals:

- Late Proterozoic   800–600 million years ago
- Ordovician and Silurian   460–430 million years ago
- Pennsylvanian and Permian   350–250 million years ago
- Late Neogene to Quaternary   4 million years ago

During these four periods, many glacial advances and retreats occurred. These advances and retreats involved continent-sized ice sheets. Scientists have proposed several physical mechanisms that

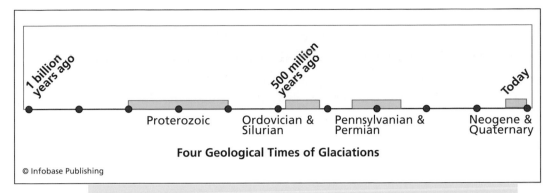

Four Geological Times of Glaciations

© Infobase Publishing

Major periods of glaciation occurred four times during the geologic past. During these intervals, there were many glacial advances and retreats. Earth is currently in a short and warm "interglacial" period.

control the onset, duration, intensity, and ending of glacial events. Included are a wide range of diverse contributions: changes in Earth's orbit, referred to as the "Milankovitch cycles"; changes in plate tectonics; reductions in *carbon dioxide* (CO_2); continental uplift; volcanic activity; solar output; and surface *albedo*. It is the combination of any and/or all of these cyclic factors and their influence on Earth's surface that trigger glacial and interglacial episodes. It is important to understand how each of these triggers relates to and impacts climate change and *global warming*.

The Milankovitch Cycles

Earth's orbit does not remain fixed; it varies over time. Major contributing factors to ice ages are these natural variations that occur between the Earth-Sun relationship. There are three natural cycles that have an effect on Earth's climate. The cycles have to do with Earth's axial tilt, its elliptical orbit around the Sun, and its "wobble," or precession (the direction the North Pole points).

These orbital parameters have cyclic occurrences (periodicities) that range from 22,000 to 100,000 years. Geologically speaking, these time frames are relatively short. Because of this, these cycles have more of an influence on the advance and retreat of ice sheets and glaciers during an ice age than they do in the timing of an ice age (which can be separated by millions of years).

These three natural variations in Earth's orbit are commonly referred to as the Milankovitch cycles, named after Milutin Milankovitch, a Serbian astrophysicist, in 1920. His theory states that variations in Earth's orbit through time cause changes in the amount and intensity of incoming solar *radiation* (insolation) that reaches Earth's surface. These differences in insolation directly affect the behavior of ice sheets and glaciers. Depending on where the cycle's periodicities are in relation to each other, this partly determines the behavior of the world's ice sheets, glaciers, and climate.

Earth's axis is tilted, a characteristic it has had since it formed. Some scientists believe Earth was struck by a massive object in space while it was being formed, knocking it off the perpendicular; others speculate the

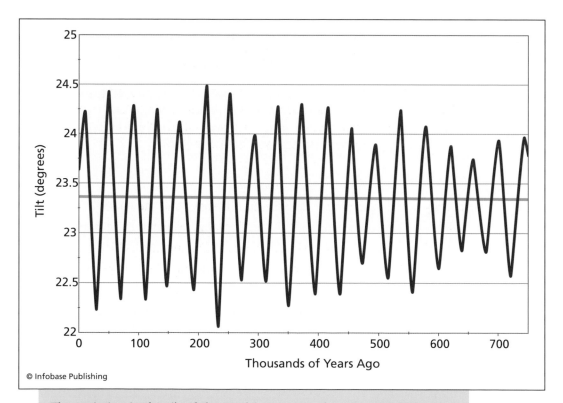

© Infobase Publishing

The variation in the tilt of the Earth's axis over the past 750,000 years: The red line represents the tilt in degrees, which varies between 22 and 24.5. The gold line represents today's value for comparison.

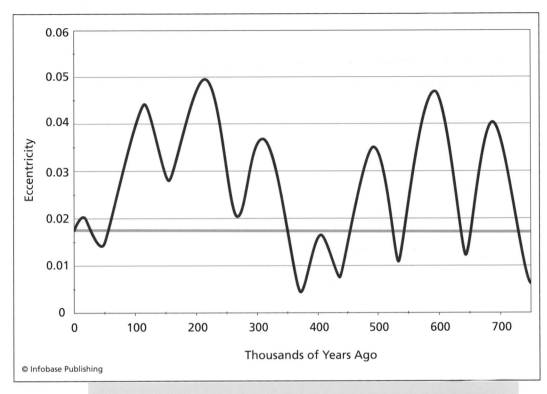

The graph shows the variation in the eccentricity of the Earth's orbit over the past 750,000 years, from circular (0.0) to elliptical (0.05). The gold line represents today's value for comparison.

tilt may have something to do with past ice ages and the distribution and weight of massive ice sheets. It is the tilt that gives Earth its seasons. Currently that tilt is approximately 23.5 degrees. During Earth's history, the tilt has varied from 21.6 to 24.5 degrees. Over the past 750,000 years, Earth's tilt has changed on a 41,000-year cycle, as shown in the figure above.

When Earth's tilt changes, the seasonal distribution of insolation at the higher (polar) latitudes and the length of long, cold winter periods at the poles also change. Changes in tilt do not have much effect on the equatorial regions. The higher the degree of tilt, the more pronounced are the seasons. The more extreme the winters are at the poles (longer and colder), the more likely ice sheets are to grow.

Earth's eccentricity is the second cycle that influences insolation. Earth's orbit around the Sun is not a perfect circle; instead, it is an ellipse, which varies from 1 to 5 percent. With a periodicity rate of approximately 100,000 years, it affects the amount of radiation Earth's surface receives at aphelion and perihelion. Aphelion is the point on its orbit where Earth is farthest from the Sun; perihelion is the point where it is the closest. This cycle changes the seasonal contrasts in the Northern and Southern Hemispheres. When the orbit is very elliptical (oval shaped), one hemisphere will have hot summers and cold winters and the other hemisphere will have warm summers and cool winters. When the orbit is more circular, both hemispheres will have similar seasons.

Precession, in the third cycle, has a periodicity of about 22,000 years. Twice each year the Sun is positioned exactly over the equa-

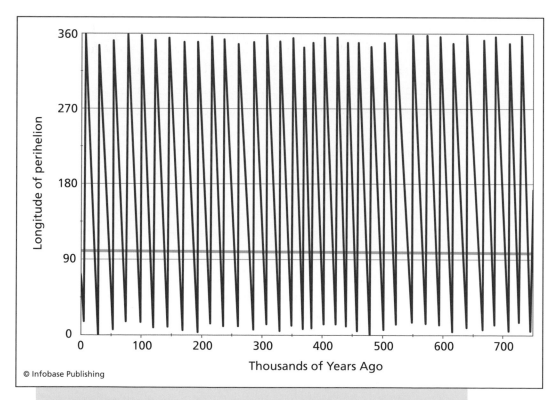

© Infobase Publishing

The precession of the equinox over the past 750,000 years: The "longitude of the perihelion" is the precession from the vernal (spring) equinox.

tor—on the two days referred to as the "equinoxes" (one in the spring and one in the fall). Presently, the equinoxes are on March 21 and September 21, but they will not always be because Earth's axis of *rotation* wobbles like a top does when it begins to wind down. As it "wobbles," the timing of the equinoxes changes. When the equinoxes change, so do the aphelion and perihelion. This affects Earth's climate because it affects the seasonal balance of insolation.

Currently, perihelion is in January for the Northern Hemisphere winter, which makes winters milder. During its 22,000-year cycle, the precession can cause significant changes in Earth's climate and in the growth and recession of ice sheets.

Plate Tectonics

Ice ages can be triggered when Earth's continents are in positions that block or reduce the flow of warm water from the equator to the poles. If warm water is not allowed to travel poleward, ice sheet formation begins. Therefore, the location of Earth's continents has had a significant influence on glaciation. Ice sheets require large landmasses at polar and high-latitude locations in order to become established and grow. These conditions have been met during Earth's ice ages. During the Ice Age (the last one), the continents of Antarctica, North America, and Eurasia supported extensive ice sheets and glaciation. In the ice age prior to that (in the Pennsylvanian and Permian), the supercontinent Pangaea was located at the South Pole and was heavily glaciated. Having continents located in close proximity to Earth's polar regions and the presence of extensive amounts of sea ice are two major factors in promoting ice ages.

Once these conditions are established, a positive *feedback* loop can begin. Because ice sheets have a high albedo (reflectivity), they greatly reduce the absorption of solar radiation because most of it is immediately reflected back into space. When less radiation is absorbed, the *atmosphere* cools down, which further promotes the growth of ice sheets. This then increases the albedo, and so on. The system becomes self-perpetuating, and the ice age continues until another mechanism occurs to break the cycle, such as volcanic activity, which increases the amount of CO_2 in the atmosphere and intensifies the *greenhouse effect*.

Lower CO_2 Levels in the Atmosphere

Just as adding carbon dioxide to the atmosphere is contributing to the warming of the atmosphere today, scientists have suggested that a lowering of atmospheric CO_2 levels can lead to a cooling of the atmosphere, pushing Earth into an ice age. Several processes can lead to a decrease in the CO_2 levels in the atmosphere, such as physical processes with ocean currents, erosion, and volcanoes.

One example is the Himalayas. Because of the extreme height of these mountains, they have increased the amount of rainfall Earth receives. This added precipitation serves to wash large amounts of CO_2 out of the atmosphere, decreasing the greenhouse effect and contributing to the promotion of ice ages. The Himalayas' formation began around 70 million years ago when the Indo-Australian plate collided with the Eurasian plate. The Himalayas are still rising today and its documented rise fits with the climate theory of the region.

Continental Uplift

Based on data collected by the U.S. Geological Survey (USGS), another suggested influence on the occurrence of ice ages is from continental uplift. According to the USGS, major uplift at continental plate boundaries can cause significant changes in oceanic and atmospheric circulation patterns. The USGS has suggested that climatic changes caused by uplift play an important role in the development of ice ages. As Earth's plates get subducted, mountains are uplifted. A well-known example is that of the Himalayas and the Tibetan plateau. Continents there have been uplifted about 2,000 feet (600 m) during the past 15 million years. This uplift is important because it can change global ocean and atmospheric circulation patterns, which in turn control the climate. For example, if warm air in circulation from the equator to the polar regions pushes up against a tall mountain range but cannot rise and pass beyond it, the warmth that the mass of air contains, which potentially could be delivered to the polar location, becomes blocked. Because the warmer air mass is physically obstructed from being delivered to the polar region, the temperatures there remain cold, potentially encouraging ice age conditions. The same is true for ocean currents. If an uplifted landmass blocks the poleward travel of a warm ocean current, the polar areas will not receive any benefit from the warm temperatures, which could trigger an ice age–like climate at the pole. It is possible that these climate changes caused by the uplift of landmasses play a significant role in the beginning of ice ages.

Volcanic Activity

Volcanic activity is another trigger of ice ages. The major volcanic gases released into the atmosphere during an eruption are water vapor (H_2O), carbon dioxide (CO_2), and sulfur dioxide (SO_2). Gases released in smaller amounts can include hydrogen sulfide (H_2S), hydrogen (H_2), carbon monoxide (CO), hydrogen chloride (HCl), hydrogen fluoride (HF), and helium (He). During a volcanic eruption, ash can reach great heights in the atmosphere and spread around Earth. This cloud of ash acts as a blanket to block out the incoming solar radiation. If the eruption is massive enough, the ash cloud can effectively block incoming solar radiation for a couple of years, causing a worldwide cooling.

According to the National Aeronautics and Space Administration (NASA), volcanic gases are thought to be responsible for the "global cooling" that is experienced after major volcanic eruptions. The degree and intensity of the cooling effect varies based on the specific characteristics of the eruption. The factors that control it include the force of the eruption, the amounts of particular gases emitted, and the geographical location of the volcano with regard to major atmospheric circulation patterns. If the eruption material also contains sulfur dioxide gas, when it reaches the *stratosphere,* it turns into sulfuric acid particles *(aerosols),* which serve to reflect the Sun's rays, causing an even greater reduction in the amount of sunlight reaching Earth's surface. This process can cause a significant cooling effect.

According to scientists at Michigan Technological University, two indices are used to measure the effects of volcanic eruptions on climate: the dust veil index (DVI) and the volcanic explosivity index (VEI). The DVI uses estimations of the amount of material dispersed into the atmosphere in addition to the temperatures at Earth's surface and the amount of sunlight reaching Earth's surface. The VEI ranks eruptions using specific criteria to determine the magnitude, intensity, dispersion, and destructiveness of a volcanic eruption. Eruptions are ranked from 1 to 8, with 8 being the most explosive. Eruptions that inject material into the stratosphere have a VEI of 4 or higher.

One of the most well-known examples of this was the eruption of Mount Pinatubo in June 1991. The eruption was so violent and ejected so much gas and particulates into the stratosphere that it temporarily offset the predicted greenhouse warming effect for a few years. During major eruptions, millions of tons of sulfur dioxide gas can reach the stratosphere. According to NASA, observations of the effects of Mount Pinatubo aerosols on global climate have been used to validate scientists' understanding of climate change and the ability to predict future climate change. In fact, researchers at NASA's Goddard Institute for Space Studies (GISS) in New York City have applied the general circulation model (GCM) of Earth's climate to it and have been successful in predicting the effects of the sulfate aerosols from Mount Pinatubo's eruption and its part in cooling global temperatures.

Short-term "global cooling" has been linked with other volcanic eruptions throughout history. Another notable occurrence was an eruption in 1816 that caused "the year without a summer." When the Tambora volcano in Indonesia erupted, it emitted 200 million tons (181 million metric tons) of sulfuric acid aerosol into the stratosphere, causing significant *weather* disruption worldwide. According to researchers at Michigan Technological University, the average decrease in temperature in the Northern Hemisphere was between 0.7° and 1.2°F (0.4°–0.7°C). Western Europe experienced weather disruptions, and the United States and Canada suffered killing summer frosts. The sulfur dioxide aerosol layer in the stratosphere caused brilliant sunsets around the world for several years afterward.

About 71,000 years ago a far more significant volcanic eruption occurred in Sumatra. Mount Toba erupted and affected the climate to the point where the next 1,000 years saw the death of many species throughout the world. In 1783, the Laki Fissure eruption occurred in Iceland and lasted for eight months. During the eruption, more than 100 million tons (91 million metric tons) of sulfur dioxide was ejected into the atmosphere. As a result, Iceland suffered great losses. Three-quarters of its livestock died, and all the agricultural crops failed. The famine that it caused led to the starvation of 25 percent of the Icelandic population.

Solar Activity

Variations in solar activity are seen as another variable in the triggering of ice ages, as the relationship between low solar (sunspot) activity and cold periods throughout Earth's history demonstrates. In the figure, the timing of low sunspot activity coincides with intervals of cooler climate on Earth. NASA released a new computer *climate model* that reinforces the long-held theory that low solar activity could have changed the circulation in Earth's atmospheric circulation in the Northern Hemisphere from the 1400s through the 1700s and triggered the Little Ice Age in regions such as Europe and North America. Change in the amount of radiation received from the Sun figures prominently into this episode. It was so cold during the Little Ice Age that transportation canals in

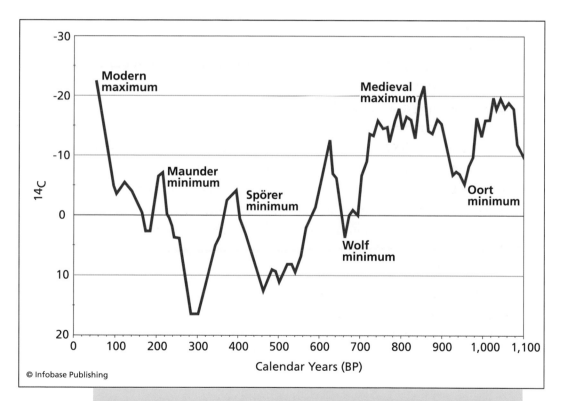

This graph represents the cyclic solar activity recorded in radiocarbon records. The Maunder minimum is associated with the Little Ice Age occurrence.

Holland were frozen, glaciers were actively flowing in the Alps, Iceland was surrounded by sea ice, and there was virtually no access to Greenland because of all the ice surrounding it.

Albedo

Albedo is another factor that has had an influence on the spread and retreat of ice sheets and glaciers. When large areas of Earth are covered by ice and snow, large amounts of insolation are reflected off the ice and snow and returned to space. This is a "high" albedo. When large amounts of the Sun's energy are returned to space instead of being used to heat Earth, it lowers the temperatures at Earth's surface. In a high-albedo situation, a positive feedback scenario can be put into motion; the cooler the area becomes, the more snowfall occurs and more ice builds up and spreads.

Conversely, if ice melts, and darker surfaces—either of land or ocean—become exposed, it can lower albedo and begin to heat up the surface of Earth. In this case, as the area heats up, more ice and snow is melted. This chain of events can continue until all the ice melts.

According to NASA, one concern scientists have today is that of ice collecting deposits of soot or other dark pollution particles on it. When this happens, the dark deposits on the ice absorb the incoming sunlight and heat up the area, melting the ice. This has caused widespread melting in several areas in the Arctic. Scientists at NASA are concerned that this could upset the delicate balance of nature in the polar regions. According to Dr. James Hansen, a climate change expert at NASA's GISS, "This provides a positive feedback—as glaciers and ice sheets melt, they tend to get even dirtier." Hansen has discovered that soot's effect on snow albedo may be contributing to trends where thinning Arctic sea ice and melting glaciers and *permafrost* may be contributing to earlier springs in the Northern Hemisphere. "Black *carbon* reduces the amount of energy reflected by snow back into space, thus heating the snow surface more than if there were no black carbon," Hansen comments.

SOURCES OF INFORMATION

In order to understand Earth's ice ages, scientists must be able to piece together the past and determine exactly when in the geologic time

This drill bit is tough enough and sharp enough to slice through meter after meter of solid ice. The bit is fitted in the core sleeve, a hollow tube into which the ice passes after being cut. *(Lonnie Thompson, Byrd Polar Research Center, Ohio State University, NOAA Paleoclimatology Paleo Slide Set)*

table these events occurred. The principal techniques they use are the analysis of ice cores and the study of oxygen isotopes.

In Antarctica, there is a 40-day time window in January and February (the Southern Hemisphere's summer) when ice core samples are generally acquired. Bubbles caught in the ice cores are representative of Earth's atmosphere at that time. The analysis of the bubbles trapped in the ice cores, as well as monitoring the movement of the ice sheets, allows scientists to look back in time to understand just how Earth looked and what processes were occurring millions of years ago.

Ice cores are a cylinder of ice four to five inches (10–13 cm) in diameter that are drilled out of the ice with a sharp drill bit. This coring device is sharp enough to slice through foot after foot of solid ice. The bit is fitted into a core sleeve, a hollow tube in which the ice is kept after it is cut to protect it.

Ice cores are one of the most accurate records climatologists have available with which to study the past. For example, a 1.9-mile (3-km)-long ice core removed from Antarctica preserved snowfall records for the past 740,000 years. According to the online site ScienceDaily researchers were able to determine past atmospheric temperatures as well as what the *concentrations* of gases and particles were in the atmosphere through time. Scientists were thus able to determine that Earth had gone through at least eight previous ice ages and interglacial periods. Dr. Eric Wolff of the British Antarctic Survey notes, "It's very exciting to see ice that fell as snow three-quarters of a million

Ice-core research is currently conducted in both Antarctica and Greenland, where climatologists study the evidence and attempt to put together the Earth's past. *(University of Wisconsin–Madison Physical Sciences Laboratory, Ice Coring and Drilling Services and National Science Foundation)*

years ago." Because the ice was deposited layer after layer on a yearly basis, interpreting the ice core is similar to interpreting the rings on a tree: Owing to the freezing-thawing differences between summer and winter, summer snow is less dense than winter snow. The resolution (detail) in the ice core is usually good enough to determine exact dates of occurrences. The principal method climatologists use to reconstruct Earth's past temperatures, which allows them to piece together ice ages in the past, is through the analysis of the isotopes of the oxygen and hydrogen atoms that make up the water molecules that initially fell on the ice sheet as snow. Isotopes are atoms that differ from one another in atomic weight. The number of protons and electrons are the same in all atoms of a particular element, but the number of corresponding neutrons can differ. For the element oxygen, there are eight protons, but there can be eight, nine, or 10 neutrons, which means it has three isotopes.

The standard oxygen atom in a water molecule is oxygen-16 (^{16}O) with two hydrogen atoms. There does exist, however, a "heavy water" combination of oxygen and hydrogen that acts the same chemically as ordinary water. The difference in the latter is that it is composed of oxygen-18 (^{18}O). Because the water component is slightly heavier, it does not evaporate as quickly as the "lighter water." Therefore, when water vapor does begin condensing, the heavier water is the first to join the water droplets or ice crystals that fall as rain or snow. During this pro-

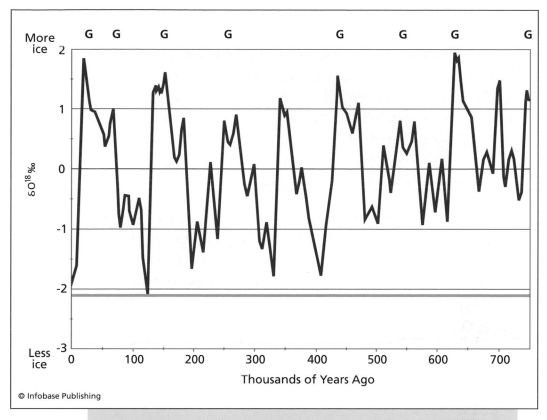

This graph shows the changing ice volume on Earth over the last 750,000 years. The different isotopes of oxygen found in foraminifera were analyzed to obtain the graph. The *Gs* designate the times when glaciers may have reached as far south as the midwestern United States.

cess, as the air becomes colder and colder, there is less and less heavy water in its water vapor.

Because of this physical property, when the number of oxygen-18 atoms in the ice or snow is compared with the number of oxygen-16 atoms in it, it can be used as a measuring stick of how cold the air was when the snow initially fell. The fewer the oxygen-18 atoms, the colder the air was. The ice core can then be dated to determine how long ago these temperatures occurred. The graph above shows the occurrence of eight major ice ages based on the $^{18}O/^{16}O$ records.

CLIMATE CYCLES

Climatologists know that Earth has naturally cycled through several ice ages in the past. Interestingly, there is a theory that at least on four different occasions, Earth has been completely held within the grip of an ice age—referred to as "snowball Earth"—only to come out of that and switch to a *tropical* state—referred to as "hothouse Earth." Snowball Earth theory was first proposed by Brian Harland, a Cambridge geologist in 1964, when he discovered glacial deposits near the equator. Later it was hypothesized by Paul Hoffman and Dan Schrag, professors in the Department of Earth and Planetary Sciences at Harvard University, when they proposed that nearly 700 million years ago Earth's climate cooled and the polar ice caps expanded. As the ice advanced, more sunlight was reflected back into space, cooling off Earth even more. The process gained momentum until Earth was completely covered in ice 0.6 mile (1 km) thick. Hoffman and Schrag have supported their hypothesis by collecting supporting evidence from sites all around the world. For example, in a lecture Hoffman delivered at the University of Texas at Austin in April 2006, he cited the following field evidence they had obtained in support of their snowball Earth hypothesis:

- carbon isotopic data that shows significant changes in the carbon content of the sea water bracketing the snowball/glaciations periods
- paleomagnetic data that depicts globally distributed glacial deposits in the mid and low latitudes occurring at the same time snowball Earth existed

(opposite page) Snowball/hothouse Earth alternations. Global temperatures fall and ice packs form, reflecting solar energy back into space. The atmosphere cools, and global temperatures plummet. The Earth becomes entombed in ice. With no rainfall, volcanic carbon dioxide accumulates in the atmosphere, and the planet begins to warm and melt the sea ice. An intense greenhouse effect begins, melting the ice, and forcing the Earth into a hothouse condition.

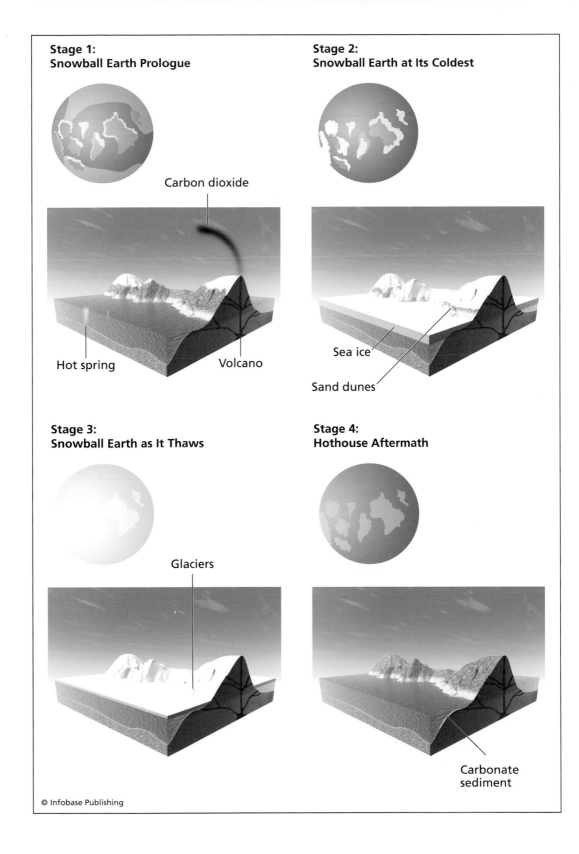

Stage 1:
Snowball Earth Prologue

Carbon dioxide

Hot spring

Volcano

Stage 2:
Snowball Earth at Its Coldest

Sea ice

Sand dunes

Stage 3:
Snowball Earth as It Thaws

Glaciers

Stage 4:
Hothouse Aftermath

Carbonate
sediment

- tidal bundles that indicate a shallow marine depositional environment and the formation of an iceline feature close to the equator
- banded iron formations, which are found only in glacial marine strata along with the existence of ice-covered continents
- deposition of postglacial "cap carbonates," which are sedimentary structures found on coastal margins, formations that exist when a "hothouse" environment occurs after a snowball Earth episode

According to scientists at the National Oceanic and Atmospheric Administration's (NOAA) Paleoclimatology Research, hothouse Earth followed snowball Earth causing a great swing in climate.

Approximately 600 million years ago, the first snowball Earth gripped the planet, burying the Tropics in ice. Even the bottoms of the oceans froze, despite the heat in Earth's molten core, because temperatures were extremely cold—down to -58°F (-50°C). All life on the planet was killed except for the most simplistic organisms. The only disturbances on the vast, dead landscape were the volcanoes spewing carbon dioxide, the gas that would slowly accumulate and melt the ice through a global warming process. Surprisingly, it took only a few centuries to put the Earth into a brutal hothouse environment, with temperatures climbing to 122°F (50°C). Many scientists believe this cycle occurred four times between 580 and 750 million years ago. The figure on page 17 illustrates how this process may have occurred.

Hoffman and Schrag have found evidence worldwide to support the existence of these occurrences, such as glacial deposits in tropical areas, mixed with iron-rich rocks that could only have formed in an environment with little or no atmosphere, topped with deposits of carbonate rock formed in warm water. The carbon levels in the rock also indicate that a scarce level of life existed through these episodes.

These ice ages were perpetuated by their high albedo to such a point that it became a runaway response. When ice formed at the tropical latitudes, Earth's albedo rose at a faster rate because the sunlight was striking a larger surface area head on, equaling a huge amount of lost energy. Once it got to this point, surface temperatures dropped rapidly, allowing the planet to freeze over.

One theory is that these episodes occurred because at the time the Sun was about 6 percent weaker than it is today. Other theories have suggested the geographic distributions of the continents played a factor. Other scientists strongly oppose the existence of snowball and hothouse Earth, arguing that if ice and snow completely covered the planet, it would have been impossible for the ice to melt and what these snowball Earth episodes warn scientists of today is that Earth is capable of experiencing abrupt climate change and the current activity of humans on the environment with global warming needs to be taken seriously.

PROMINENT COOLING EVENTS

One of the most studied ice ages is that of the last one, when 20,000 years ago the global temperature was 9°F (5°C) lower than it is today and wooly mammoths and saber-toothed tigers roamed the present-day New York City area. During this period, commonly known as the Ice Age, one-third of the planet's land surface was covered with ice. In North America, an imposing ice sheet up to two miles (3.2 km) thick covered Canada and reached as far south as southern Illinois. Ice sheets also covered Greenland, northern Asia, and northern Europe.

Scientists, from professionals at the USGS to glacial geomorphologists at universities, study the landforms left behind to piece the trail of ice advances together. They also deduce past conditions through the analysis of biological evidence. Fossil remains have been discovered of species such as reindeer and wolverines—cold climate species—in areas that today would be much to warm and inhospitable for their survival. By radiometrically dating these specimens, climatologists and paleontologists, such as those at the USGS and NOAA, can determine when the colder climate conditions existed. Examples of this include musk ox fossils that have been found as far south as Mexico. Mastodon fossils have been located off the coast of present-day New Jersey in the Atlantic Ocean; walrus once existed off the coast of present-day Virginia.

The La Brea tar pits in Los Angeles are another area where Ice Age life has been preserved and has been radiometrically dated, giving scientists at the Natural History Museum of Los Angeles and from universities useful data for reconstructing past climate. This type of proxy evidence (using related physical phenomena as an indicator of something else) offers proof that during the Ice Age, conditions were cooler much farther south, and ocean levels were significantly lower.

Since the Ice Age, there have been some prominent cooling events. One of the most well known is that of the Younger Dryas event, which occurred around 11,000 years ago. The Younger Dryas was a brief but intense return to cold climate conditions, almost leading to reglaciation, and lasted for about 1,000 years. This weather event serves as a good example of how rapidly climate can oscillate. Its beginning was abrupt; in just 20 years atmospheric temperature dropped 12°F (7°C). Just as dramatically, at the end of the thousand years, the atmospheric temperature returned to its previous state.

Initially, there were two theories that suggested its cause: a major influx of icebergs deposited into the ocean from a disintegrating Arctic ice shelf or shifts in the patterns of orographic (mountain) winds in response to the retreat of the ice sheets. These theories had major flaws, however, because they could not explain how the event could influence such a large geographical area and how it could start and stop so rapidly. A subsequent proposal for the cause was that a diversion of the meltwater occurred between the Mississippi and St. Lawrence Rivers. This theory was developed by Wallace Broecker of the Lamont-Doherty Earth Observatory in New York. The meltwater coming off of the massive Laurentide Ice Sheet drained down the Mississippi and into the Gulf of Mexico. Then, through a sequence of events, it was gradually diverted to eastern outlets via the St. Lawrence. The last of these "eastern diversions" documented was when the Lake Agassiz drainage was routed across Canada into the St. Lawrence estuary. At this critical point, all the drainage was now shifted to the east. The eventual impact of this was to shut down the deepwater circulation in the North Atlantic. Based on this model, as glaciers were retreating, their meltwater sank to the bottom of the ocean and triggered the "conveyor belt" of oceanic circulation. Because the oceanic conveyor system serves to transport heat to the north, it caused the remaining ice sheets to melt faster. The Younger Dryas was first documented in northern Europe, through the discovery of proxy evidence found in marine cores, pollen samples, and terrestrial plants. There are some problems with this model, as well. It does not adequately explain the same effect in the Pacific Ocean or the rapid fluctuations that occurred in Greenland. Because of this, the "Heinrich events" have also been suggested as part of the cause.

The Heinrich events were caused by a multitude of icebergs calving and entering the Atlantic. Iceberg calving is the process by which pieces of ice break away from the terminus of a glacier that ends in a body of water or from the edge of a floating ice shelf that ends in the ocean. Once they enter the water, the pieces are called icebergs. This would have caused a sudden sea-level rise of many feet (m) and lowered the elevation of the Laurentide Ice Sheet. This would then have caused the Northern Hemisphere to warm. Increased snow accumulation would have allowed the ice sheet to return to its former elevation and begin a new cold phase called the Younger Dryas. The extent of the Younger Dryas was global, not just regional.

Another sudden decrease in global temperature occurred around 8,200 years ago (circa 6200 B.C.E.) and lasted 200–400 years. This cold interval, however, was not as severe as the Younger Dryas. Scientists have discovered evidence for this cooling phase in ice cores retrieved from Greenland and in sedimentary records from the temporal and tropical regions in the Atlantic. This episode was global in nature and affected sea level worldwide.

It is possible that this episode was caused by the introduction of massive amounts of meltwater still flowing off the Laurentide Ice Sheet of northeastern North America and flowing into the North Atlantic. Another possibility was the sudden draining into the North Atlantic of glacial Lake Ojibway. This huge influx of fresh meltwater once again negatively impacted the *Gulf Stream* and the thermohaline circulation of heat transfer in the North Atlantic, causing an abrupt climatic cooling. In *temperate* zones, temperatures dropped 9–11°F (5–6°C); in the Tropics, 5°F (3°C).

Conditions became cooler and drier; documented evidence also exists that the changes in climate were severe enough that they impacted early human settlements. Areas in East Africa experienced 500 years of drought as a result. This climatic condition lasted approximately 400 years before climate reverted back to normal.

The next notable cooling episode is referred to as the "Little Ice Age," which lasted from about 1350 to 1850 C.E. This was characterized by the advance of glaciers worldwide and unusually cold winters in North America, Europe, and Asia. Scientists suggest the cause for this cold interval was low solar activity and an increase in volcanism. From 1645 to 1715, there was a documented interval of low solar activity referred to as the *Maunder minimum*. It has been suggested that the episode was directly linked to the cold temperatures of the Little Ice Age. The Earth's volcanoes were extremely active during this time period, also, and some believe it is possible that the addition of ash and sulfuric acid particles to the atmosphere may have contributed to the lower temperatures.

The cooling during this episode was not as extreme as in previous ones—roughly only 1.67°F (1°C). It is also not clear that the effects were global. Cold winters and the advance of glaciers show up in records for Europe and North America, where agricultural practices were negatively impacted. Around 1850, the Little Ice Age came to an end, possibly owing to human activity and global warming.

THE HOCKEY STICK THEORY

The hockey stick theory refers to a study conducted by Dr. Michael Mann of the Department of Geosciences at the University of Massachusetts and his associates Raymond S. Bradley and Malcolm K. Hughes. Using tree-ring proxy data, he reconstructed global climate for the past 1,000 years. The resulting graph resembles a hockey stick: The majority of the time period represents little change in temperature and would-be the "shaft" of the stick: a sharp upward rise in the 20th-century portion of the graph illustrates sudden temperature rise and constitutes the stick's "blade." When constructing the graph, Mann used tree-ring data for the pre-1900 portion of the graph and used surface temperature records for the 20th-century portion. The visual effect was very dramatic, illustrating the fact that the 20th century's climate was rising out of control. The year 1998 was especially problematic. This year—which also suffered an extreme El Niño event—was noted as the "warmest year of the millennium." A problem with this, pointed out later, was that data collected via *satellite* did not exactly correlate to Mann's. In the meantime, this single paleoclimate study became the "foundation" of the global warming theory. The hockey stick graph was adopted by the Intergovernmental Panel on Climate Change *(IPCC)* and used as a strong basis for evidence as to the urgency of recognizing global warming and the need to take action immediately.

In 2003, Stephen McIntyre, a Canadian mineral exploration consultant, and Ross McKitrick, an economist at the University of Guelph, in Ontario, challenged Mann's theory, accusing him of using inappropriate data, methodology, and statistical methods. McIntyre and McKitrick charged Mann with creating the hockey stick with a "collation of errors, incorrect calculation, and other quality control defects." They claimed that the wrong places, the wrong dates, and the wrong numbers were jumbled together to produce the results Mann wanted: proof that the *Industrial Revolution* had started global warming. Further, they accused Mann of using only 12 sets of proxy data, obtained only from the Northern Hemisphere. He then extrapolated that data to reach the conclusion that global temperatures remained stable and then dramatically increased in the 20th century.

Shortly thereafter, Mann and his associates recognized some slight data preparation errors that had been made and printed a correction. The hockey stick graph still existed, however, so even with minor errors in the original statistical tests, it was not off far enough to change the trend of the upswing in temperature. But the controversy over global warming science had only just begun.

The dispute led to an investigation by the U.S. Congress at the request of Representative Joe Barton of Texas. A panel of scientists convened by the National Research Council (NRC) of the National Academy of Sciences and chaired by Edward Wegman, a statistician, stirred up the scientific community and the issue became embroiled in hot debate.

After many studies and most people's minds made up, there is still a lot of controversy around global warming. The basic results currently stand that although McIntyre and McKitrick raised many sensitive issues, their claims have been rebutted in detail by many other independent climatologists around the world. Meanwhile, McIntyre and McKitrick are still objecting. The majority of climate scientists worldwide support the hockey stick theory.

Hans von Storch, the director of the Institute for Coastal Research of the GKSS Research Center in Germany, agrees with McIntyre and McKitrick that Mann's methodology was not completely sound, but he also added that correcting it would not alter the overall results: The 20th century was still warmer, as the hockey stick showed.

The bottom line with this study, however, is that the controversy is based on a single paleoclimatic study that took place eight years ago. Since then, dozens of temperature reconstructions have been created in order to better understand climate change worldwide. Although many of these studies do not show the shaft part of the hockey stick as straight (there are a few ups and down in it), the same message is very clear: The 20th century is noticeably warmer than the rest of the millennium, and the 1990s were likely warmer than any other time in that period. This falls in line with exactly what Mann has said all along, as well as the IPCC and thousands of scientists worldwide.

WILL THERE BE ANOTHER ICE AGE?

Although it is not possible to predict climate, scientists are very aware of the capacity of climate to change abruptly. In light of all the attention today on global warming, many people may think a future ice age is out of the question as long as the "enhanced greenhouse effect"—that being consciously added to by humans—persists. Climates can, and have, changed in unexpected ways. The Younger Dryas, for example, has caused much scientific debate over its cause. No known specific natural "forcing" has been identified for its occurrence. Actually, the event occurred at a time when orbital forcing should have caused the climate to continue to become warmer.

The one phenomenon that scientists come back to over and over when climates change drastically and unexpectedly is "ocean dynamics," or more to the point—thermohaline circulation. Even though Earth's surface temperatures are rising rapidly today, ice age effects are not impossible. If large masses of ice and glaciers are melted and their freshwater is drained into the North Atlantic, it can stop the ocean circulation that supplies heat from the equator to the polar areas, thereby cooling Earth significantly. This type of abrupt climate change is discussed in detail in chapter 6.

CHANGES TODAY BECAUSE OF WARMER TEMPERATURES

Global warming is a worldwide phenomenon, and although some areas will be impacted to a greater degree than others, effects will be felt everywhere. Each diverse and unique *ecosystem* will experience changes to its environment. Some areas are already feeling the effects of global warming, others may not feel them for a while. The following list outlines some of the changes occurring today:

- The Larsen B Ice Shelf in Antarctica has lost volume as large chunks (one was as large as the entire state of Rhode Island) have broken off the continent and melted.
- The surface area of Arctic sea ice has declined 8 percent over the past few decades.
- Plants and animals are already changing their habitat ranges, moving to new areas in order to survive.
- Significant increases in ocean temperatures have been measured and documented over the past five decades.
- Mount Kilimanjaro's glacier, which has existed for more than 11,000 years, will disappear by 2020 if the present rates of melting continue.
- Siberian peat lands are beginning to thaw. They are now releasing stored carbon dioxide and *methane* gases into the atmosphere, both of which are significant *greenhouse gases.*

These present changes, in turn, could have some of the following effects in the not-so-far-off future:

- Sea level could rise 7–23 inches (18–50 cm) by 2100.
- More than 1 million species face extinction from disappearing habitat.
- Arctic ice is quickly disappearing. The Arctic could be ice-free by 2040.
- Ocean corals are being hit hard. Some areas are seeing 70 percent mortality rates.

The polar regions are expected to be impacted to the largest degree. Because polar areas are so sensitive to climate change, they also serve as an early warning signal for the rest of the world.

(Source: *National Geographic,* Union of Concerned Scientists)

THE PHENOMENON CALLED GLOBAL DIMMING

Global dimming is a fairly new concept, one introduced largely to the general public at the beginning of the 21st century, although it had already existed in the scientific community. It was first documented by Gerry Stanhill, an English scientist. While in Israel, he was comparing sunlight records from the 1950s with current ones. He discovered that today there is a noticeable drop in solar radiation—up to 22 percent. And the effect is global; it is not related to just one geographic location.

Stanhill named the phenomenon "global dimming" and wrote his findings in 2001. He received skeptical responses from other scientists. Recently, however, Australian scientists used a completely different method to estimate the incoming solar radiation, and their results confirmed Stanhill's findings. After this, other scientists worldwide began taking the global dimming concept seriously.

Dimming is a result of air pollution. The burning of *fossil fuels* (coal, oil, and gas) and wood produces not only carbon dioxide but tiny airborne particles of soot, ash, sulfur compounds, and other pollutants. The visible air pollution reflects sunlight back into space, preventing it from reaching the surface. The idea is that through the introduction into the atmosphere of pollution particulates and aerosols that incoming solar radiation is either absorbed or reflected back into space. Through global dimming, the amount of sunlight reaching Earth is reduced, lower temperatures are reached, and the warming effects of greenhouse gases are masked.

Renowned climate scientist James E. Hansen, from NASA's GISS, estimates that global dimming has cooled Earth by more than 1.8°F (1°C) over the last 100 years. He is concerned that as global pollution levels are lowered, that global warming may increase to the point of no return. According to Hansen, scientists had long known that pollution particles reflected some sunlight, but they are only now realizing the magnitude of the effect. "It's occurred over a long time period," Hansen explains, "so it's not something that perhaps, jumps out at you as a person in the street. But it's a large effect."

Although improving the quality of the air is a desirable goal, part of what pollution is providing, ironically, is a counterbalance for increasing global warming. Its negative effects are masking, or offsetting, global warming's negative effects. This has created a serious problem. While pollution issues cannot be ignored—for the health of all life on the planet, it needs to be reduced—at the same time, humans need to increase their efforts at fighting global warming. Many people think global warming is really a problem for future generations to face, but this is simply not true. Because of pollution, global warming is actually already more serious than the public thinks, and the effects will escalate as pollution levels drop in the future. As scientists continue to study past ice ages and look at today's climate phenomena, however, it enables them to apply the principles they learn to better plan for the future.

Glacial Retreat and Meltdown

Glaciers are dynamic formations, always changing and highly sensitive to the natural input from Earth's climate. Because of their dynamic nature, they are very useful "measuring sticks," used by climatologists and other scientists to identify changes in the environment. Worldwide glacial retreat and meltdown is one of the most telling markers of global warming. When Earth's surface temperature begins to rise, these massive rivers of ice begin to melt and shrink in response. This chapter focuses on these highly responsive systems. It first takes a look at glacial morphology, in other words, what they are, how they move, why they move, and why they form where they do. Next, it explores the distinct landforms and signatures upon the terrain that long-gone glaciers have left, enabling climatologists to piece together the climate history of an area. The chapter then examines how glaciers reflect climate change and what the current evidence is for global warming. In conclusion, it addresses the impacts of glacial retreat and how computer modeling techniques are being used today to better understand the delicate balance between global warming and glacial retreat.

GLACIAL MORPHOLOGY

Glaciers consist of snow that has accumulated year after year. Glaciers have a positive mass balance: The accumulation of precipitation (usually in the form of snow) is greater over a year's time than the melting, leaving the glacier with a net increase in mass. As the snow deposit increases in thickness, the heavy snow burden presses down on the layers of snow accumulated beneath it. The extreme compression forces the snow to recrystallize. The grains of ice become similar in size and shape to table sugar. Over time, the grains grow larger. The pressure forces the air pockets within the mass to shrink. This process increases the ice's density and eventually compresses the lower levels into large, thick ice masses.

After approximately two years, the snow turns into a state called "firn," an intermediate state between snow and glacier ice. As the years pass, the ice crystals become so compressed and dense that any air pockets that are left are minuscule. The ice crystals in extremely old ice can grow to several inches in length. Once the snow undergoes this metamorphosis into ice, a process that can take a century or more, it becomes a glacier. This extreme density and compression is what gives glacial ice its distinctive pale blue color.

Because glaciers are a collection of a formidable, enormous mass of ice, when they form in a canyon on an inclined slope, they have the ability to move. Glaciers can vary in size from small alpine glaciers smaller that a football field to mammoth ones stretching across the terrain for almost 100 miles (160 km).

Glaciers cover roughly 10 percent of Earth's surface. As a testament to the last ice age, most of the world's glaciers are located in the polar environments of Antarctica and Greenland. Each year, these two locations attract research scientists eager to unlock the climatic secrets from the world's past. For instance, the University of Nebraska–Lincoln sent seven people to Antarctica to join a team of world-class international geoscientists for a three-month expedition to investigate the continent's role in global climate change. They are part of the ANDRILL geological drilling project, organized to recover rock-core samples from the McMurdo Sound region to study the history of the Ross Sea area's ice sheets. As the edges of the ice sheet break free and fall into the ocean, it opens the way for the interior glaciers to flow faster toward the edges of the continent.

The USGS is currently involved in mapping the changes in area and volume of the polar ice sheets in both Antarctica and Greenland and relating the impact of recent melting to worldwide rising sea levels. In other studies, NASA is busy measuring the shape of the ice sheets and determining whether they are getting thicker or thinner in different areas and what they may mean for the short- and long-term, such as advancement or retreat. They are trying to model the interior glacial "mass balances."

In Greenland and the Arctic, scientists are also trying to understand the mass balance of glacial ice. Konrad Steffen, a glaciologist from the University of Colorado, has camped on Greenland's ice sheet every year since 1990. His goal is to determine the way that meltwater affects ice movement. He lowered a camera 330 feet (100 m) into an ice crevasse to determine if the huge waterways under the ice could be seen. He discovered that as more water channels through these underground tunnels, the glacier can shift more rapidly.

The U.S. National Science Foundation (NSF) is particularly heavily involved in Greenland and the Greenland Ice Sheet Project Two (GISP2). This scientific endeavor is to understand environmental change in the Arctic in order to help current science better predict future situations and needs. Meanwhile, a new study, led by Son Nghiem of NASA's Jet Propulsion Laboratory, has used buoys and satellite imagery to prove winds since 2000 have pushed huge amounts of thick ice out of the Arctic basin past Greenland.

The National Snow and Ice Data Center at the University of Colorado, funded by NASA, has been involved in monitoring the extent of Arctic sea ice at the end of each summer. Josefino Comiso, who is a senior researcher for NASA's Goddard Space Flight Center announced that the ice is melting at an accelerated rate. The seasons when sea ice melts, between early spring and late fall, have become much longer and warmer each decade.

One of a glacier's unique characteristics is that it can move. Movement occurs as soon as a glacier becomes about 60 feet (18 m) thick. Under the imposing mass of the ice, coupled with Earth's gravity, glaciers move as an extremely slow river of ice. They can move down canyons, across flat areas, and out to sea. According to the climatic conditions, glaciers can rapidly or slowly advance or retreat, providing an indication of climatic trends. Typically, movement occurs over a long period of time, but occasionally, glaciers have been known to surge forward up to 33 feet (10 m) per day over a short interval of time.

Global warming, however, is causing many of Earth's glaciers to retreat. Rapid glacial retreat can be visible over just a few months or years. When glaciers are traveling down mountain canyons, they can join together where canyon junctions merge. As the mass of the glacier grinds over the ground, it breaks up pieces of rock and soil and carries the debris

This mountain glacier is located in the East Coast Mountains of Baffin Island, Nunavut, Canada. This river of ice flows downhill at several feet per year. Glaciers begin in high mountain peaks and flow down steep mountain valleys, where they eventually merge into one another and flow as a giant river of ice, eroding the mountainsides and valley bottom as they grind their way downhill. *(John T. Andrews. NOAA Paleoclimatology Program/Department of Commerce; INSTAAR and Department of Geological Sciences, University of Colorado, Boulder)*

Tidewater glaciers are the glaciers that calve icebergs into the ocean. They can pose a problem for ships that navigate nearby ocean waters. *(NOAA)*

along the bottom and sides of the glacier. This debris being carried along takes on a banded appearance running the length of the glacier.

The bulk of the world's glaciers are in Antarctica and Greenland, but they also occur on other continents. Because they must meet specific geographic and physiographic criteria in order to survive, they only occur in specific areas. Glaciers form above the snow line on mountains because these are areas that receive large amounts of snowfall during the winter season and experience cool temperatures in the summer. These conditions must exist so that the snow and ice do not melt from one season to another.

There are several types of glaciers: mountain glaciers, hanging glaciers, piedmont glaciers, tidewater glaciers, cirque glaciers, and valley glaciers. Mountain glaciers are those that develop high in mountainous areas. They can occupy a single mountain peak or be spread across an

Types of Glaciers	
GLACIER TYPE	CHARACTERISTICS
Hanging glacier	• Attached to steep mountainsides • Looks wider than they are long • Forms in steep, mountainous areas, such as the Alps in Europe
Piedmont glacier	• Forms when a steep valley glacier flows out onto a broad, flat area where it spreads out into a wide fan shape
Tidewater glacier	• Flows across land until it reaches the ocean • Calves icebergs that are common in Arctic waters
Cirque glacier	• Forms in a bowl-shaped hollow on a mountainside
Valley glacier	• Originates from mountain glaciers that flow downhill and are channeled into valleys • Can extend for many miles • Often flows far enough to reach coastal areas, where it empties into the ocean

entire mountain range. The most common areas to find mountain glaciers are in the Arctic regions of Canada and Alaska, the Himalayas in Asia, the Alps in Europe, Antarctica, and the Andes in South America.

PAST GLACIAL EVIDENCE

Glacial erosion has significantly reshaped Earth's surface with several distinct types of erosional and depositional features over geologic time. When climatologists and other scientists study the landscape and find these features, they are able to unravel clues about the area's past climate and determine when glaciers existed in the area and what the climate was like at that time.

HOW ICE MELTS

For years, scientists believed that when ice began to melt, the actual process was fairly simple: When a solid begins to heat up, the molecules within the ice simply acquire more energy and start jiggling around. The more they jiggle, the faster they transform from their solid state to a liquid. Based on a report from LiveScience.com, scientists now know that while this is basically true, it is not the entire explanation; there is more to it. Before this process begins, another, far subtler, occurrence happens. The melting actually begins as the atomic structure begins to "crack." This process is so minuscule that scientists have not been able to actually see the process.

In order to solve this issue, Arjun G. Yodh, professor of physics and astronomy at the University of Pennsylvania, created a way to conduct the experiment on a larger scale using transparent crystals resembling small beads. This allowed him to view the process in an optical microscope. The crystals behaved like a huge version of the atoms. Yodh and colleagues were able to determine that a "premelting" process occurs in the area where the atoms within solid crystals are not perfectly aligned, and they begin moving. Acting as imperfections, they begin moving first, then spread to the ordered portions of the crystals. This discovery illustrated that when ice melts, there is a "premelting" process that occurs before the actual melting temperature is reached.

One very interesting study was conducted by Dr. Julian Dowdeswell, director of the Scott Polar Research Institute at the University of Cambridge. Dowdeswell, along with a research team from Cambridge, was the first to use images taken of sediments found beneath the ocean floor to locate and identify glacial landforms from ancient glacial activity. Their methodology has offered an entirely new approach to paleoclimatology. In their study, seismic data and detailed images provided a three-dimensional view of the ocean floor in the area around the mid-Norwegian continental shelf. The research team was able to "see" deposits left approximately 2.5 million years ago, marking the first attempt at seeing 3-D ancient glacial landforms.

What they were able to find were elongated and streamlined ridges of sediment deposited by fast-moving ice streams. There were also plow marks carved out by the bottoms of icebergs being dragged along the seafloor. Ridges and moraines were also deposited transverse to the ice flow as the ice retreated (in a similar fashion as to how they are on land today).

As Dowdeswell has observed, "Submarine landforms like these are found at the marine margins of almost all modern ice masses, so it seems reasonable to assume they were also features of ancient ice sheets." He also adds, "What no one has done until now, however, is search for signs of these glacial landforms from millions of years ago. New developments in geophysics are giving us the ability to produce detailed images from hundreds of metres beneath the modern seafloor, and seismic data that enable us to produce 3-dimensional pictures of what the surface of Earth was like when it was being shaped by glaciers. That opens up new, exciting possibilities in identifying past ice ages—which we intend to take further."

Glaciers erode the landscape by the processes of abrasion and plucking with the sheer mass of the glacier traveling over the ground. The ice is able literally to lift the weaker blocks of material up and drag them along with the ice flow. This happens when water at the base of the glacier flows into fractures in the ground's surface. Once water enters, it freezes. When water freezes, it expands, which widens the fractures, eroding and weakening them further.

This process weakens the rock so that it can be plucked out of the ground and carried in the bottom of the glacier as it flows downhill. This "bedload" that the glacier constantly acquires as it flows abrades the ground as it moves, grinding rock into a flourlike substance, and leaving deep scratches in the ground's surface that geologists can later use to identify the existence of a glacier and the direction it was moving. When glaciers flow down mountain valleys, they leave a characteristic U-shaped valley.

As a glacier moves, it can carry an enormous load of broken pieces of rock from the mountain further up canyon. These pieces of broken rock can vary in size from small stones and dirt up to rocks the size of cars and be deposited on hillsides. These deposits are called glacial

One of the most recognizable landforms left after the existence of a glacier is the U-shaped valley formed from the abrasive grinding of the ice as it plows its way downhill. *(Nature's Images)*

erratics. Once the ice thins and melts, it can no longer carry all this heavy load of material, so it dumps it where it is. Moraines are another common landform feature. These are formed from the deposition of material from the glacier that is exposed once the glacier has melted and retreated. They are a mixture of all the eroded material the glacier collected and carried along with it as it traveled. Moraine deposits are common along the edges of where the glacier once flowed. They are also common at the terminus, or furthest reach, of the glacier.

Glaciers also deposit snakelike ridges called eskers that are formed by streambeds underneath the glaciers, as well as drumlins, which are streamlined hills. Another common glacial deposit is loess (pronounced "luss"). Loess is a very fine sediment, often referred to as "rock flour," and formed from the extreme abrasive action of glaciers. It is easily picked up by wind and blown over long distances to be deposited elsewhere. The loess deposits can be very deep; some in China are hun-

Erratics are a common glacial deposit. These granite boulders deposited on the hillside vary in size. The large one on the upper left is larger than a car. The granite pieces were carried about 10 miles (16 km) downhill during the last ice age about 12,000 years ago. *(Nature's Images)*

dreds of feet deep. They also occur in the Midwest of the United States, principally the Mississippi Valley and Great Plains regions. There are also smaller deposits in Idaho and Washington and more significant ones in Alaska.

HOW GLACIERS REFLECT CLIMATE CHANGE

Glaciers serve as one of the most valuable proxy tools in which to study climate change because they can range in age from hundreds to several thousands of years old. When a core is extracted from a glacier, the bubbles are analyzed in it, just as they are in ice cores. Analysis of the oxygen content reveals what the composition of the atmosphere was like at the time that layer of snow fell on the glacier. In addition, analyzing the $^{18}O/^{16}O$ ratio reveals what the temperature was like. Vegetation

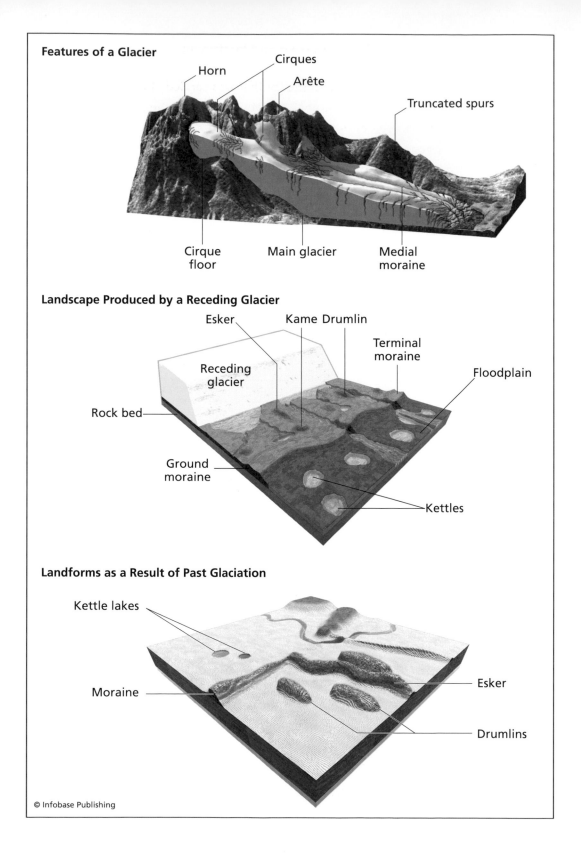

Features of a Glacier

Horn

Cirques

Arête

Truncated spurs

Cirque floor

Main glacier

Medial moraine

Landscape Produced by a Receding Glacier

Esker

Kame Drumlin

Terminal moraine

Floodplain

Receding glacier

Rock bed

Ground moraine

Kettles

Landforms as a Result of Past Glaciation

Kettle lakes

Moraine

Esker

Drumlins

© Infobase Publishing

and other debris trapped in the ice that was carried along with the glacier can also be used to unlock secrets about what the past climate was like. Dust reveals information about the wind conditions, salt content can give evidence of windiness near oceans, and sulphuric acid content relays information about nearby volcanic activity.

As scientists such as Dowdeswell and paleoclimatologists at the NSF, USGS, National Ice Core Laboratory, NASA, and NOAA Paleoclimatology Research Center trace Earth's history back from one ice age to another, they are able to build a temperature chronology and determine by how many degrees the planet warmed up during each interglacial period and cooled off during each glacial interval. It is because glaciers are so dynamic and responsive to temperature change that they are good indicators today of temperature changes occurring worldwide in the face of global warming. The fact that so many of the world's glaciers are currently melting reflects a change taking place in the climate right now. The following lists what some of the world's glaciers are currently experiencing:

- In the Peruvian Andes, Qori Kalis Glacier is losing as much ice in one week as it used to lose during an entire year.
- The Alaskan glaciers are losing an average of six feet (1.8 m) of thickness each year.
- The glaciers in the European Alps are expected to be gone by the end of the 21st century.
- Mount Kilimanjaro's glacier in Africa is expected to be gone in about a decade.
- The glaciers that supply drinking water to northern India and western China will no longer be able to provide this life-sustaining service in just a few years.
- The glaciers in Canada's and Montana's Glacier National Park will all be melted in the next few decades.

(opposite page) There are several diagnostic features of a glacial landscape that climatologists use to identify areas that were glaciated in the past. Recognizing these features gives them insight as to what the climate was once like.

EVIDENCE OF GLOBAL WARMING

Since the early 1900s, most of the glaciers around the world have been melting and retreating at rates faster than have ever been recorded. Since the beginning of the Industrial Revolution in the mid-1700s, scientists have noted that glaciers, which are extremely responsive to current temperature change, have been receding. Changes will continue to happen as long as the atmosphere continues to heat up. According to the IPCC, "Ample evidence indicates that global warming is causing glaciers to retreat worldwide." Changes can be observed, measured, and monitored in order to determine how fast glaciers are melting and how that impacts humans. Today's technology has also stepped in. Space satellites are now able to measure the global temperature trend. Mass balance of a glacier is calculated (mass added each year compared to mass melted) as cubic meters of water lost, or as thickness averaged over the entire area of the glacier. The findings are sobering. The world's sources of freshwater are being reduced. Areas of the world that depend on melting glaciers for their drinking water are suffering because this natural resource is becoming scarcer. As an example, the Himalayan glaciers feed into seven of the great rivers of Asia and supply year-round water to 2 billion people. In the Ganga (one of the rivers that receives water from the Himalayan glaciers), the loss of glacier meltwater would cause water shortages for 500 million people and for 37 percent of India's irrigated land. Countries in South America are faced with similar situations.

In other areas where glaciers are melting rapidly, flooding is another threat that is being measured. Sometimes huge chunks of glacial ice can fall into lakes and cause glacial lake outburst floods, killing many people. Another consequence that is being measured is sea-level rise. As ice sheets' meltwater enters the ocean, it raises the water level, flooding coastal areas. As more glaciers continue to melt, coastal towns will be flooded and destroyed.

The Arctic, maybe more so than any other place on Earth, has felt the effects of global warming. Its fragile ecosystem is very vulnerable to change. It is especially susceptible to changes in climate. One reason for its fragility is that it is a relatively new ecosystem; up to as recently as 20,000 years ago it was covered by glaciers in the ice age, and the ecosystem, therefore, did not exist. *Weathering* is also a very slow process in colder climates, which means that in the Arctic it takes an extraordinary amount of time for soil to form, as compared to the hot, humid tropical areas on Earth. The arctic soil is also very low in nutrients, putting the plants at a disadvantage. Two other conditions plant life has against it are that it must live in conditions of low temperature and low light. The Arctic ecosystems are truly the most extreme conditions found on Earth, and any life that survives must learn to adapt to these extreme conditions. Because of all these disadvantages, Arctic ecosystems have an overall lower productivity compared to other ecosystems. These natural hardships that ecosystems must face in the Arctic also puts hardships on the wildlife that live within the habitat. When situations such as global warming and pollution enter this fragile environment, it can be especially devastating, when compared to other ecosystems, simply because it does not have as much resistance. Change in the Arctic, however, means change everywhere. Because polar areas play an important role in Earth's heat balance, what affects the Arctic, ultimately affects the entire planet.

Scientists have identified several pieces of evidence that global warming is occurring. For example, the 10 warmest years ever recorded all occurred within the past 14 years. Ice cover at both poles is shrinking. The Greenland Ice Sheet has undergone a record amount of melting, and glaciers in Canada and Alaska are disappearing at record rates. The Arctic growing season has lengthened by a few days per decade over the years. Permafrost has started to thaw, causing many problems for inhabitants of those polar areas, as the ground shifts and settles.

If the snow and ice begin melting in the Arctic, they will no longer be able to reflect incoming sunlight, causing the planet eventually to heat up. The Arctic can also cause a rapid global warming if the frozen peat bogs begin thawing and large amounts of carbon dioxide (CO_2) are released into the atmosphere.

According to proxy records in ice cores, peat cores, and lake sediments, Arctic air temperatures have risen by 0.8°F (0.5°C) each decade over the past 30 years. Most of this warming has occurred in the winter and spring. In addition to this, atmospheric temperatures from the late

1900s to the present are at their highest level in the past 400 years. Climatologists have documented that Arctic and Northern Hemisphere river and lake ice have been forming later and melting earlier over the last few centuries. In fact, since 1850, the total ice-cover season is now 16 days shorter. According to the Yukon Conservation Society, sea ice in the Arctic covers about 10 to 15 percent less area in the spring and summer than it did in the 1950s. The ice is also estimated to be about 40 percent thinner in the late summer and early fall than it was in recent decades. When the darker water around the ice edges starts absorbing heat, the ice at the edges of the open water quickly melts away and the darker area gets bigger. This larger area absorbs more heat. This starts a cycle that soon speeds up. In the Arctic Ocean, as global warming causes warmer temperatures, more ice will melt because the air is warmer. In addition, satellite technology has confirmed that each year the number of snow-free days increases in the Arctic and the growing season extends by a few days.

One of the most obvious pieces of evidence that climatologists have keyed in on is the multitude of shrinking glaciers over the past 50 years. Even more alarming, during the past 10 years, the melting rate has become three times faster than what it was before. In a cause-and-effect relationship, this means that melting glaciers are contributing to sea-level rise. According to the USGS in 1986, there had been a warming at ground level of 3.3 to 6.7°F (2–4°C) during the past few decades. Thawing permafrost is disastrous on developed areas with highways, roads, and other structures. At present, the USGS is actively engaged in research concerning the emission of methane hydrates (a greenhouse gas) from the permafrost layers as it melts. This could contribute significantly to global warming. Currently, they have researchers participating in research projects in Alaska, Canada, and Russia, studying and monitoring environmental impacts and implications. They are also interested in melting sea ice and what impact that is having on global sea rise.

GLACIAL RETREAT

Glaciers today are retreating worldwide, not only in the polar regions, but also in the mid-latitude mountain ranges such as the Rocky Mountains, Himalayas, Alps, Andes, Cascades, and Mount Kilimanjaro. They are retreating faster than what can be explained by historical rates. Because this glacial retreat became more pronounced after 1850, it is more likely correlated to *anthropogenic* (human-caused) actions. With the further progression of the industrial revolution, extensive use of fossil fuels, and practices such as *deforestation* and certain farming techniques, the atmospheric temperature has steadily risen. Because glaciers are so delicately balanced, the effects of these warming temperatures are being seen first hand by their response.

When a glacier reaches the point where snow accumulation is less than ablation (loss of ice), the glacier cannot maintain its mass balance and will begin to retreat. Climate change and global warming can affect this both in terms of temperature and snowfall. Unfortunately, today there are very few advancing glaciers left on Earth. The following passages discuss glacial conditions in various geographic regions.

Arctic Regions

According to the World Wildlife Fund (WWF), melting occurring in the Arctic has accelerated in the late 1990s. Current estimates of combined melting have increased from 39 square miles (100 km²) per year 1980–89 to 124 square miles (320 km²) in 1997 and 208 square miles (540 km²) in 1998. Today, the glaciers in Greenland are some of the fastest moving in the world. In fact, they are contributing to about one-sixth of the current annual rise in sea level. Containing 12 percent of the world's ice, over the course of the past five years, their rate of flow has doubled, accelerating to a rate of eight miles (13 km) a year.

Areas in the interior of Greenland are increasing in mass from current precipitation, but there has been significant thinning and loss of ice around the periphery (edges) of the landmass. The edges are melting quickly, and huge chunks of the Greenland Ice Sheet are sliding into the ocean. A suggested reason for this is that as the surface melts, water seeps to the base of the ice and lubricates it so it slides faster. Up until now, scientists did not think this could happen so fast; rather, they thought the process took hundreds to thousands of years.

According to Eric Rignot of NASA's Jet Propulsion Laboratory, in 1996, Greenland poured 90 times more water into the sea than Los

Angeles consumed, and in 2006, it poured 225 times more. In 2016, he predicts, Greenland will be pouring 450 times more. In an interview with zFacts, he stated: "The speedup in Greenland has been detected simultaneously in many glaciers. When you have this widespread behavior of the glaciers, where they all speed up, it's clearly a climate signal. The fact that this has been going on now more than 10 years in southern Greenland suggests this is not a short-lived phenomenon."

Dowdeswell, of the Scott Polar Research Institute, believes that the accelerated outflow of glaciers in Greenland has been caused by two important triggers: the breakup of ice tongues that reach all the way out into the sea at the glaciers' leading edges, which has a similar effect as uncorking a bottle—remove the stopper and the glaciers begin to readily flow—and the melting at a record rate during the past few years of ice on the glaciers' surface. Melted water that percolates down through crevasses in the ice reaches the base of the glacier, lubricates the sediments on the ground, and allows the ice to slide. What concerns scientists is that adding meltwater to the Atlantic Ocean could inhibit the flow of the currents in the North Atlantic, shutting down the current that supplies the tropical heat to the polar areas. Oceanographers have reported that some components of these currents have already shown evidence of slowing by 30 percent since 1992.

On a more positive note, although most of the Arctic glaciers have been melting the past few decades, some located in Scandinavia and Iceland have actually increased because they have seen an increase in precipitation in the past few years.

Antarctica

According to the National Snow and Ice Data Center, in 2006, Antarctica contained approximately 95 percent of all the world's freshwater *resources*. While the frigid temperatures keep the surface below freezing temperature, scientists have discovered that the bottoms of the glaciers at the places where land meets the ocean are melting rapidly throughout the entire continent. Rising ocean temperatures due to global warming have been suggested as the cause. Warmer ocean temperatures are also believed to be responsible for the current accelerated thinning and breakup of the multitude of huge, floating ice shelves, which have also caused the accelerated glacial flow toward the ocean.

After the Larsen B Ice Shelf collapsed in 2002, the glaciers nearby have notably accelerated in their flow and are still speeding up. Since the Wordie Ice Shelf broke in 1995, many major ice streams that feed the Larsen A shelf are flowing two to three times faster toward the ocean now. Crane Glacier increased its speed from 5.6 feet/day (1.7m/d) to 10.2 feet/day (3.1m/d) in April through December of 2002, and then to 13.5 feet/day (4.1 m/d) between December 2002 and February 2003—a nearly 250 percent increase in speed. The ice shelf has acted as a barrier, slowing the glacier's flow. This is a concern, because if larger ice shelves should ever collapse—such as the Ross Ice Shelf—it could trigger an outpouring of several glaciers into the ocean.

The interior of Antarctica is receiving an increase of snow accumulation, however. Scientists believe this is because the water that is being evaporated from the warmer ocean water surrounding the continent is condensing and falling as snow on the interior of the continent. Therefore, right now this process is offsetting some of the melting, but scientists do not know how long that will continue.

Europe

According to research conducted by the WWF in the European Alps since 1980, melting has accelerated; in the past two decades, glaciers there have decreased in size up to 20 percent. In the Alps of Switzerland, it has been reported that 84 of the 85 surveyed glaciers are retreating; only one is advancing. Others in Sweden and Norway are also rapidly retreating.

In one particular study in Norway, a group of research scientists from Swansea University, in the United Kingdom, launched the field study project Sea Level Rise from Ice in Svalbard (SLICES), designed to measure and calculate past and future sea-level rise. The team used aerial photographs of the glaciers in Norway and built highly detailed 3-D digital elevation computer models of the Svalbard glaciers.

In their analysis of Slakbreen Glacier, they were able to determine the glacier's surface had lowered 328 feet (100 m); and since 1961, the snout (front) had retreated 0.9 mile (1.4 km). Most of this retreat had occurred in the last 10 years. The group of researchers believes that global warming is a cause for concern to the general public. While small glaciers like these represent only 4 percent of the world's total land ice, they account for roughly 20 to 30 percent of 20th-century sea-level rise, and this melt has greatly increased since 1988.

Using high-tech computer modeling techniques, the SLICES team was able to forecast sea-level rise for the 21st century under different climate conditions. They view their work as critical to help those who live along the world's coastlines today. According to the World Meteorological Organization, the summer 2003 temperatures that triggered landslides, floods, and the formation of glacial lakes were some of the hottest temperatures ever recorded in the northern and central regions of Europe. Scientists warn that if these warming trends continue, the glaciers in the European Alps will virtually disappear over the next few decades.

Asia

The WWF has reported in its studies in 2005 that 67 percent of all Himalayan glaciers are retreating. Out of 612 glaciers located in China since 1990, 95 percent are reported as retreating. The Bhutan Himalayas are currently retreating 98–131 feet (30–40 m) per year. In Kazakhstan, glaciers have been shrinking, losing 0.8 square mile (2 km^2) of ice each year since 1953. The Chinese Meteorological Administration believes the country's northwestern mountains will lose 25 percent of its glaciers by 2050. This presents a critical problem for China, because these glaciers supply 15 to 20 percent of the drinking water for more than 20 million people; it is an extremely valuable resource the nation cannot do without.

All of the glaciers in the Mount Everest region are currently retreating. In fact, Khumbu Glacier, which lies on the main route to Everest, has retreated 3.1 miles (5 km) since 1953.

South Pacific

In Indonesia, 80 percent of glacial mass disappeared between 1942 and 2000. According to the WWF, West Meren Glacier, in particular, receded 8,530 feet (2,600 m) from 1936 until it finally melted sometime

Muir Glacier, in Alaska. The top image was taken in 1941; the bottom, in 2004. *(National Snow and Ice Data Center. Top photo by William O. Field; bottom photo by Bruce F. Molnia)*

Shepard Glacier in Glacier National Park, Montana, has melted significantly over a 92-year period due to rising temperatures. The top photo was taken in 1913; the bottom, in 2005. *(Top photo by W. C. Alden, NPS Archives; bottom photo by Blase Reardon, USGS)*

Grinnell Glacier from Mt. Gould
1938 - 2006

1938
Hileman photo
GNP Archives

1981
Key photo
USGS

1998
Fagre photo
USGS

2006
Holzer photo
USGS

Grinnell Glacier in Glacier National Park, Montana *(USGS)*

around 1997. In the South Pacific, mountain glaciers in New Zealand have been retreating since 1890. In recent years, the melting rate has significantly sped up. In particular, both the Fox and Franz Josef Glaciers are 1.6 miles (2.5 km) shorter than they were 100 years ago. Of the 127 glaciers surveyed in New Zealand's Southern Alps, they have all shortened by nearly 40 percent and lost 25 percent of their area since the mid-1850s. On a more favorable note, however, some of these have advanced recently due to more precipitation in the area.

North America

The North American glaciers located along the Rocky Mountains in the United States and Canada and the Pacific Coast Ranges from northern California to Alaska are nearly all in a state of active retreat. Since 1980, the rate of retreat has increased each decade.

In the North Cascade Mountains alone (part of the Pacific Coast Ranges), there are more than 700 glaciers; the area extends from central Washington to the Canadian border. These glaciers store a tremendous amount of water, equal to all the lakes and reservoirs in Washington State alone. These glaciers are a critical resource for inhabitants of the area because they provide water each summer. Unfortunately, since the mid-1980s, the North Cascade glaciers have lost an average of 41 feet (12.5 m) in thickness and nearly a quarter to half of their volume.

Of the 47 glaciers in the North Cascades that are being monitored, all are receding. Even worse, Lewis Glacier, Spider Glacier, Milk Lake Glacier, and David Glacier, have completely melted and disappeared.

In Glacier National Park in Montana, glaciers have also been disappearing rapidly. In 1850, the park was a showcase of 150 unique glaciers. Today, it has only 26 left. In fact, more than two-thirds of its glaciers and about 75 percent of its area covered by glaciers have completely disappeared. The National Park Service and the USGS have mapped the area of each glacier in the park for decades, and in 1997, the USGS began the Repeat Photography Project to monitor and document the physical retreat of the glaciers in the park. Glaciers throughout the park are retreating and being documented. The larger glaciers are only about one-third of their former size.

Dr. Daniel Fagre, a federal research scientist based at Glacier National Park, has stated that "because of global warming, . . . estimates [are] that all the glaciers in the park will be gone by the year 2030." He explains, "The glaciers respond to global warming by having less snow in the winter and then they start melting earlier in the spring. The summers, of course, are always a period when they melt but these are now longer. It would take a pretty substantial climate change to bring our glaciers back. We would have to get a lot of moisture and it would have to get cooler. So I don't think that we'll see much of a change in the next few decades."

In the Canadian Rockies, an outlet glacier called the Athabasca has retreated 4,921 feet (1,500 m) since the late 1800s. Some glaciers in Canada stopped retreating and had brief periods of increase for a few years but then switched back into a long-term trend of recessions.

There are thousands of glaciers in Alaska. Columbia Glacier, near Valdez in Prince William Sound, has retreated 9.3 miles (15 km) during the past 25 years. Mendenhall Glacier near Juneau, Alaska, has retreated 1,902 feet (580 m). The only glacier that has shown a positive trend is Taku Glacier.

Africa

According to the WWF, the tropical glaciers in Africa have decreased by 60 to 70 percent since the early 1900s. On Mount Kenya, seven of the 18 glaciers that existed in 1900 had completely melted by 1993, and some of the remaining glaciers have currently lost 60 to 92 percent of their area. Glaciers on the Ugandan side of the Rwenzori Mountains of the Uganda-Congo border are also melting rapidly.

A significant glacier in peril is the one on Mount Kilimanjaro in Tanzania. Since 1912, the glacier on the mountain's summit at 19,340 feet (5,895 m) has retreated 75 percent. The volume of glacial ice is 80

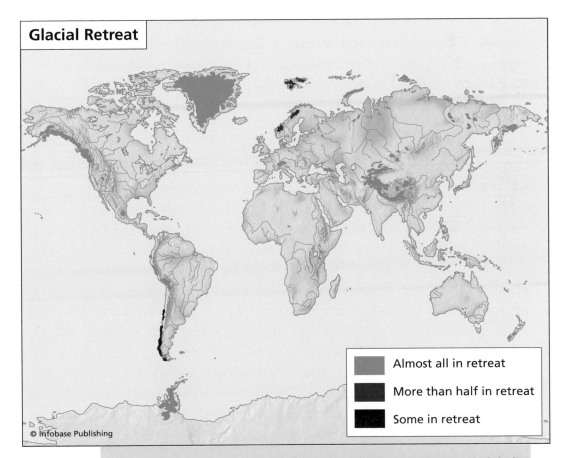

Glacial Retreat

Almost all in retreat

More than half in retreat

Some in retreat

© Infobase Publishing

This map shows the locations of the world's melting glaciers. If global warming continues, some day these glaciers may be gone forever.

percent less than it was only a century ago. If the current warming trend continues, Kilimanjaro's glacier, which has existed for more than 10,000 years, will disappear by 2020.

South America

In South America, the Northern Andes have the largest concentration of glaciers located in the tropical regions, but they are melting rapidly, especially since 1990. In Ecuador, the Antizana Glacier melted eight times faster during the 1990s than it did in the previous decades. Glaciers in Bolivia lost almost half of their area and two-thirds of their volume during the mid-1990s. The Chacaltaya Glacier could disappear by 2011–2016. According to WWF research, the Northern Patagonian Ice Field in Chile and the Southern Patagonian Ice Field in Chile and Argentina had lost only 4 to 6 percent of their 1945 area by the mid-1990s, but thinning has accelerated recently. In fact, some areas in the southern ice field had thinning rates from 1995 to 2000 that doubled the rates of the previous 30 years.

IMPACTS OF GLACIAL RETREAT

The melting and retreat of glaciers have the potential to impact millions of people. These impacts will be felt worldwide, although some areas will be affected more than others. The effects will involve drinking water and irrigation resources—something that everyone in the world depends on—the generation of electrical energy, the health of habitats connected to glacial regions, the presence and use of biochemical elements, flooding, and resulting sea-level rise. Each of these issues will have an effect on the environment in which we live.

Drinking Water and Irrigation

Many communities depend on glacial meltwater each spring and summer for their drinking water. In tropical regions, the glaciers melt year-round, supplying drinking water to people and animals that would not have any other source of water for survival. According to the WWF, the Himalayan glaciers alone provide a year-round supply of water to more than 2 billion people. In the Ganga, if glaciers further shrink, up to 500 million people may suffer from a shortage of drinking water. There are multitudes of people in Ecuador, Bolivia, and Peru who also rely heavily on glacial meltwater as their source of drinking water. In addition to drinking water, glacial meltwater is also frequently used to irrigate crops. Areas such as South America and Central Asia rely heavily on glacial meltwater for these purposes.

Electricity

Glacial meltwater is also used to fill up reservoirs, which serve not only as a source of drinking water but also to generate hydroelectric power. If glaciers retreat to the point where their meltwater can no longer keep the reservoir filled and the lake level gets too low, they will lose the ability to produce electricity. This will affect a significant number of households that depend on that glacial meltwater to generate the energy to carry on their daily activities.

Ecosystem Health

When glacial melt changes and slows down, it also impacts the ecosystems along the stream channels. When stream flow rates and sea-level rates change, habitats can be seriously impacted. Animals and organisms that live near or on the ice are also affected. According to the WWF, global warming has already caused the loss of an entire ecosystem on the ice shelves of the Arctic. Between 2000 and 2002, the Ward Hunt Ice Shelf off Ellesmere Island in Canada broke in two and emptied Disraeli Fjord. This was a 3,000-year-old lake that supported an ecosystem hosting rare microscopic organisms at the bottom of the lake. Scientists reported that by 2002, 96 percent of the unique habitat had been lost.

Biohazards

The WWF reports another impact from melting ice in the form of contaminants. Before organic pollutants such as PCBs and DDT were banned, they were widely used during the mid-1900s. PCBs are polychlorinated biphenyls, a chemical mixture used in the past as coolant and as insulating fluid for transformers and capacitors. They are no longer used in the United States, but they still can be found in the environment. DDT stands for dichlorodiphenyltrichloroethane, which was a chemical pesticide used in the past but was also banned in the United

States. A significant amount of the long-lasting pollutants were transported by air and deposited on the glacial ice, where they have been stored all this time, locked in the ice. As the ice melts, these pollutants are being let loose into the atmosphere.

Flooding

Another impact of melting glaciers is flooding. When melting occurs too rapidly, river banks can reach full capacity and overflow their banks. They can also form glacial meltwater lakes. If these lakes become too full, they can burst and cause a catastrophic flood in villages downstream. In fact, according to the United Nations Environment Programme (UNEP), there are currently 44 glacial lakes in Nepal and Bhutan alone that are in immediate danger of overflowing as a result of global warming. Specialists at the WWF report that in Peru, a huge chunk of glacier ice is ready to fall into a lake. If this should happen, it could cause a flood that could hurt or kill 100,000 people.

Sea-Level Rise

If glaciers continue to melt at their current rate, they will add 0.008 to 0.02 inch (0.2–0.4 mm) to sea levels. It is possible, however, that this amount may be greatly underestimated. It has been recently discovered that the melting rates of Alaska's glaciers and the Northern and Southern Patagonian Ice Fields have greatly increased and are now adding an additional 0.01 inch (0.03 cm) each year. If glacial melt causes sea levels to rise, it will hurt coastal regions worldwide by increasing flooding and erosion, allowing salt water to enter aquifers and contaminate them. It will and also enter freshwater habitats, negatively impacting the life that is supported in these biomes. According to the WWF, the small sea-level rise experienced during the 1900s led to erosion and the loss of 39 square miles (100 km²) of wetlands per year in the Mississippi Delta.

COMPUTER MODELING APPLICATIONS

A project supported by the Global Change Research Program of the USGS and conducted by the Northern Rocky Mountain Science Center at West Glacier Field Station took place in 2003 to model climate-induced glacier change in Glacier National Park from 1850 to 2100. The researchers created a 3-D computer model of the Blackfoot-Jackson Glacier Basin in Glacier National Park, Montana. Based on field evidence, the glacier had already decreased in area from 8.3 square miles (21.6 km²) in 1850 to 2.9 square miles (7.4 km²) in 1979. Daniel B. Fagre and Myrna H. P. Hall of the Northern Rocky Mountain Science Center at West Glacier Field Station confirmed that global temperatures had also increased by about 0.75°F (0.45°C) over this same time period.

The team created models that could reflect the creation and ablation (melting) of the glacial ice as well as the response of the surrounding vegetation to change. Once they had set these features up as their base level, they modeled the area's distribution of glaciers under two possible climate scenarios of the future: a carbon dioxide–induced global warming scenario and a simple linear extrapolation.

The results of the model based on global warming (Model 1) concluded that all the glaciers would disappear by the year 2030, even though there were some predictions in increased precipitation (snow) to offset the melting rate. Under Model 2, the melting rate was slower.

The research team then developed another model to analyze how the natural vegetation in the area would respond to various soil moisture levels and increasing atmospheric temperatures. They wanted to see where, in a fragile alpine landscape, the established plant communities would likely migrate to as the biologic and physiographic conditions began to change around them.

Their global warming model increased the CO_2 in the atmospheric temperature 4–5°F (2–3°C) by the year 2050. The model also calculated that precipitation (mostly rain now) would increase by 10 percent. The models showed that by 2100, long after the glacial ice had disappeared, the vegetation evolved as it moved uphill toward cooler elevations.

Hall and Fagre concluded that high-elevation trees will become established above the current tree line, which will reduce the diversity of herbaceous plants. Stream temperatures will also warm up, negatively impacting temperature-sensitive aquatic invertebrates. Other ecosystem changes such as altered soil moisture, altered fire frequency, and forest growth will also be impacted, based on their conclusions.

Glacial retreat remains one of the environment's most viable indicators of global warming and climate change. Scientists will continue to monitor them worldwide as people continue to use fossil fuels and temperatures continue to climb. They will also continue to warn world populations of negative impacts and to search for solutions to this very serious problem.

The Cryosphere and Isostasy

Global warming has already caused temperatures to rise in the Arctic to the point that ice melt is becoming a serious issue. Although it may seem that the polar regions are so remote that there is not much concern for changes in these vast, barren, hostile areas on Earth's surface, just the opposite is true. It is these areas that serve as trigger points—points where if change begins to occur in global warming, points where crucial thresholds are reached, they may become unstoppable for the rest of the planet. This chapter looks at this world of ice, the cryosphere: ice sheets, ice shelves, and ice caps. It also looks at the role they play on Earth's continents during glacial periods—a concept called glacial isostasy—and the roles all of these components play on sea level. What will continued global warming do to them? Why do scientists carefully watch the frozen portions of Earth to predict what will happen elsewhere under the impact of global warming?

THE WATER CYCLE—EARTH'S WATER STORED IN ICE AND SNOW

Earth's complete water cycle is very dynamic. The water cycle moves through its various states on or through Earth—gas, liquid, and solid—sometimes quickly, other times slowly. It moves most dynamically through the atmosphere as a gas (water vapor) and as a liquid through the atmosphere or on Earth as rain and in rivers. Water can transfer fairly rapidly through *evaporation,* condensation, clouds, and precipitation; but it can also be stored for extremely long periods of time in various reservoirs such as lakes, ponds, and catchment basins. It can be stored in its liquid state in the ocean or in deep aquifers under the ground; it can also be stored for even longer periods of time in its solid form, as snow or ice in ice caps, sheets, and shelves.

Even though the water cycle is dynamic—it is always in motion—more water is in storage at any given time than is actively moving through the system. Most of this long-term storage is in the form of ice. Nearly all of Earth's ice storage exists in two geographic locations: Antarctica and Greenland. These two locations have enormous ice caps because of the landmasses' extreme polar (high-latitude) locations. The presence of ice is also a factor of cold ocean currents, atmospheric circulation, temperature, and positive feedback mechanisms. For example, ice has a high albedo, causing most of the incoming solar radiation to be reflected back to space instead of absorbed and thereby ensuring that the ice does not melt. In fact, Antarctica contains almost 90 percent of this total, Greenland, nearly 10 percent. Of all Earth's water, the amount locked up as ice only accounts for approximately 1.7 percent. While this may not seem a significant amount, it is because it represents a huge percentage of the world's freshwater, that which is available as potential drinking water for humans. The reason why it is considered potential drinking water is because the water held in storage in ice does not contain salt, as ocean water does. According to the USGS, the water stored in ice represents around 68.7 percent of all the freshwater located on Earth. This makes the world's ice a very important natural resource. In addition, one of the largest impacts this stored ice has on the world's weather is its albedo. The high reflectance

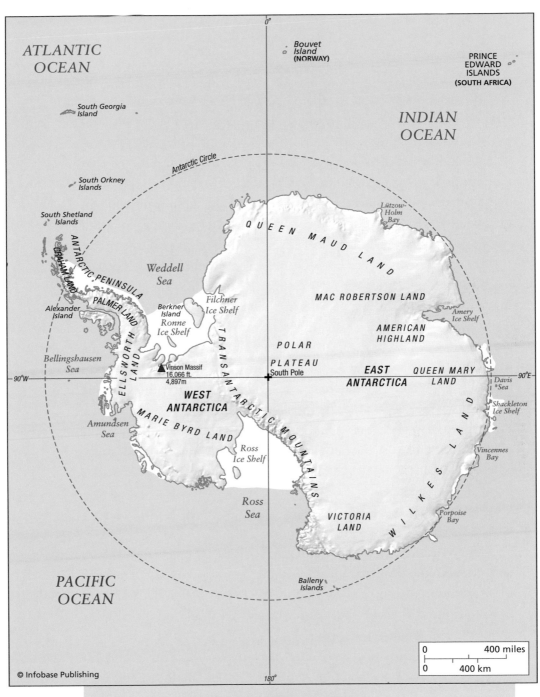

ATLANTIC
OCEAN

Bouvet
Island
(NORWAY)

PRINCE
EDWARD
ISLANDS
(SOUTH AFRICA)

South Georgia
Island

INDIAN
OCEAN

Antarctic Circle

South Orkney
Islands

Lützow-
Holm
Bay

South Shetland
Islands

QUEEN MAUD LAND

Weddell
Sea

ANTARCTIC PENINSULA
GRAHAM LAND

MAC ROBERTSON LAND

PALMER LAND

Filchner
Ice Shelf

Berkner
Island

Alexander
Island

Ronne
Ice Shelf

AMERICAN
HIGHLAND

Amery
Ice Shelf

POLAR

Bellingshausen
Sea

ELLSWORTH LAND

PLATEAU
South Pole

TRANSANTARCTIC MOUNTAINS

EAST
ANTARCTICA

QUEEN MARY
LAND

90°W

Vinson Massif
16,066 ft.
4,897m

90°E

Davis
Sea

WEST
ANTARCTICA

Shackleton
Ice Shelf

WILKES LAND

Amundsen
Sea

MARIE BYRD LAND

Ross
Ice Shelf

Vincennes
Bay

Ross
Sea

VICTORIA
LAND

Porpoise
Bay

PACIFIC
OCEAN

Balleny
Islands

© Infobase Publishing

180°

| 0 | 400 miles |
| 0 | 400 km |

Most of the world's ice is located in Antarctica.

of the Sun's incoming energy affects both Earth's temperature and wind patterns.

ICE SHELVES AND ICE SHEETS

An ice sheet is a mass of ice that covers at least 19,305 square miles (50,000 km²). In the past, ice sheets have covered different areas of Earth. Evidence of these ice sheets has been found in glacial deposits and landforms where they once existed. During the last ice age, the Laurentide Ice Sheet existed in North America, extending southward over Canada and the northern United States. During the same time, in Europe, the Weichselian Ice Sheet covered northern Europe. In South America, the Patagonian Ice Sheet covered much of that continent.

The primary difference between an ice sheet and an ice shelf is size; ice sheets are much larger. An ice sheet is larger than 19,305 square miles (50,000 km²) in area and forms a continuous cover of ice and snow over a land surface, spreading outward in all directions. An ice shelf is smaller and is attached to the land along polar coasts on one side, but most of it is afloat on the ocean surface. Because snow and ice have accumulated for a long time, ice sheets are extremely thick. While frozen on the surface, they are generally warmer at their base, heated by Earth's geothermal energy. When the geothermal heat melts the ice in contact with Earth's surface, it allows the ice sheet to flow. In the areas where this occurs, distinct channels of ice actually flow in what scientists call "ice streams."

Antarctica

The Antarctic Ice Sheet is the single largest mass of ice on Earth. It covers an area 5.4 million square miles (14 million km²), and its mass contains 11.58 million cubic miles (30 million km³) of ice. If global warming were to cause this entire sheet of ice to melt, it would raise sea levels about 215 feet (66 m).

According to a study conducted by Isabella Velicogna and John Wahr, research scientists at the University of Colorado at Boulder's Cooperative Institute for Research in Environmental Sciences, and published in *Science* in March 2006, each year the Antarctic Ice Sheet is losing up to 36 cubic miles (150 km³) of its mass—a trend that scientists are

The Antarctic Ice Sheet is the single largest mass of ice on Earth. Scientists from many countries visit Antarctica every year to study the continent's ice cores in an effort to understand the Earth's past and in hopes of better understanding the Earth's climate today and in the future. *(NOAA)*

attributing to global warming. The pair used data from NASA's Gravity Recovery and Climate Experiment (GRACE) twin satellites to conduct their analysis. As Velicogna has noted, "The ice sheet is losing mass at a significant rate. It's a good indicator of how the climate is changing. It tells us we have to pay attention."

Richard Alley, a glaciologist from Pennsylvania State University, has commented: "It looks like the ice sheets are ahead of schedule in terms of melting. That's a wake-up call. We better figure out what's going on." If this trend continues, the global sea level could rise significantly over the next several hundred years. At present, there is already so much water melting from the ice sheet and flowing into the ocean, that it has caught the attention of scientists from many countries around the world. As a comparison, each year as much meltwater flows off Antarctica into the ocean as the entire population of the United States uses in just three months. This enormous amount of water is causing the global sea level to rise 0.02 inch (0.05 cm) a year.

Some have compared Antarctica to a "slumbering giant that is being awakened." For thousands of years since the Ice Age, the ice sheet remained stable, having thinned only 985–1,150 feet (300–350 m) in the past 13,000 years. The retreat of the ice sheet during this period of time was caused by rising sea levels, not by changes in precipitation or temperature. Today's global warming poses a new situation where dramatic rises in sea level could occur and the world's ice sheets could be in jeopardy.

When the ice streams of Antarctica reach the coast and push out across the ocean, they have to flow over rocky, rough terrain. When they do this, they attach themselves to the irregular rocks along the edge of the landmass, which tend to behave like a hinge. Once the ice is "anchored," the ice continues to grow outward onto the open water. This process is what creates the huge ice shelves that are associated with Antarctica.

Currently, ice shelves cover over half of Antarctica's coastline. In fact, the ice shelves there are so significant that they equal an area about one-tenth the size of the Antarctic continent. Some of the best-known shelves are the Ross Ice Shelf and the Larsen A and B Ice Shelves. The largest is the Ross Ice Shelf in West Antarctica. It covers an area larger than the size of California and is fed by more than seven ice streams.

The ice shelves behave like a floating dock; they float up and down with the tides. As they do, they grind and grate against the rocks, slowly cracking the ice apart, forming crevasses. Because of this constant destructive process, pieces of ice shelves are constantly breaking off—calving—into icebergs. Most calving takes place during the brief period of seasonal warming. Because ice shelves are already floating in the water, they do not raise sea level when they break off. As they melt, however, sea level not only begins to rise, but also ice flows faster to the ocean that was originally held back on the land, kept in place by the ice shelf blocking the way.

Generally, the icebergs are not extremely large, and usually, the total mass of the calvings is roughly equal to the amount of new snow that falls on the continent during the year. Over the past several thousand years, this process has kept the mass balance in equilibrium.

Unfortunately, for the past few years, this has not been the case. Due to global warming, the ice shelves on the Antarctic Peninsula have been

Ice shelves are attached to the continent's shore but float on the water. They can be extremely thick. When they calve icebergs, they do not contribute to sea-level rise as they are already floating in the water. *(NOAA)*

rapidly melting. Several of the major shelves have experienced significant collapse. The Wordie Ice Shelf, for example, collapsed in the late 1980s. The Prince Gustav and Larsen A Ice Shelves collapsed in 1995. At the end of the 1990s, scientists at the National Snow and Ice Data Center (NSIDC), NOAA, and the University of Colorado all predicted that the Larsen B, Wilkins, and George VI Ice Shelves were at the "point of no return" and that the Larsen B Ice Shelf would be the next to disintegrate. They determined that the Larsen B had lost its structural strength during the summer of 1998, when an iceberg 75 square miles (200 km²) in size broke off of it. The Larsen B is huge: Its surface area is 4,800 square miles (12,000 km²). Reaching a thickness of 722 feet (220 m), it has probably existed since the last ice age 12,000 years ago. It was predicted that when it disintegrated, it would dump more ice into the Southern Ocean than all of the icebergs calved between 1950 and 2000.

Predictions rang true in January 2002. Through the use of a highly specialized satellite imaging system called the Moderate Resolution Imaging Spectroradiometer (MODIS), the shattering of a huge piece of the shelf into the ocean was monitored. It sent thousands of icebergs adrift into the Weddell Sea. The shelf did not dissolve all at once, however; the process occurred over a 35-day period, January 31–March 5, 2002. In total, about 1,255 square miles (3,250 km²) of ice disintegrated. The volume of ice that broke free measured 720 billion tons (653 billion metric tons). According to NSIDC, over the past five years, the Larsen B shelf has lost 2,201 square miles (5,700 km²) of ice, leaving it presently with only about 40 percent of its original mass. This total loss represents an area larger than the state of Delaware; the loss in 2002 alone, larger than the state of Rhode Island.

This occurrence is significant. It is the largest single event in a series of ice shelf retreats in the Antarctic Peninsula over the past three decades. Climatologists at NSIDC, which is part of the Cooperative Institute for Research in Environmental Science at the University of Colorado at Boulder, claim the retreats are due to a significant warming of the South Pole climate. By their calculations, the region has warmed 0.8°F (0.5°C) each decade since 1940. Because of this, the extent of seven of the existing ice shelves have shrunk 5,212 square miles (13,500 km²) since the mid-1970s.

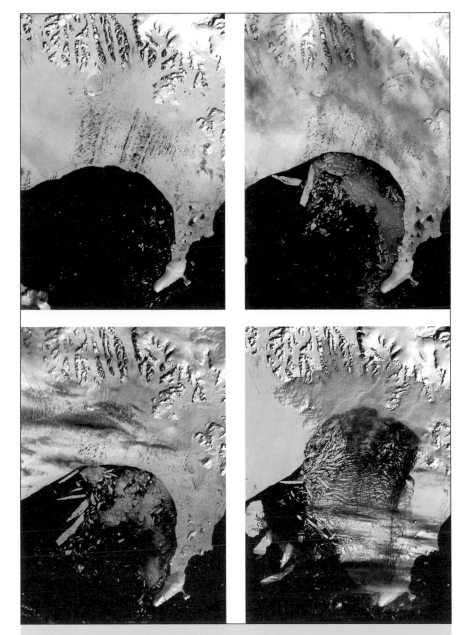

A large portion of the Larsen B Ice Shelf disintegrated in 2002, as evidenced in these MODIS satellite images from January 31, 2002 (*upper left*); February 17, 2002 (*upper right*); February 23, 2002 (*lower left*); and March 5, 2002 (*lower right*). Earth scientists predict that if global warming continues, incidences like this will become more common. *(NSIDC)*

These scientists believe there is a process in place that promotes the disintegration. They believe that melted pond water that sits on the ice's surface in the late summer contributes to fracturing the shelf because it percolates into the ice and fills the cracks. The weight of the water forces the ice apart. Researchers at NSIDC have been able to identify this process from satellite imagery.

The meltwater theory explains the disintegration of the Larsen B shelf breakup shown in the photos on page 65. In late January 2002 (summer in Antarctica), there was indeed extensive meltwater ponding on the shelf. That summer experienced unusually warm temperatures and an extended melting season. In February, satellite imagery showed several of the meltwater ponds disappear; scientists believed that they drained through open fractures into the ice. Once the fracturing reached a certain threshold, the ice sheet began breaking apart. The area of the shelf that was destroyed coincided exactly with the area that was covered by melt ponds in late January.

Scientists from 30 countries, including the United States, Britain, Taiwan, Russia, Finland, Ukraine, Japan, Argentina, Chile, Australia, Brazil, France, Italy, and Germany, conduct research in Antarctica, looking for answers to the delicate natural balance there and the effect global warming is having on the ecosystem. They are also looking for other mechanisms that contribute to ice breakup. Three other ideas have been introduced:

1. Meltwater that seeps between the ice crystals warms the shelf, where it then significantly reduces its strength.
2. Meltwater seeps into shallow cracks, expands the cracks as it refreezes during the winter, and slowly pries the ice apart.
3. Ocean warming and subice currents drag against the underside of the ice, then pull at the ice and weaken it.

Even though the melt of ice shelves does not threaten the rise of sea level, what is of concern is that the ice shelves act as a block to many inland glaciers and keep them from freely flowing into open water. If the ice shelves are gone, there is nothing to hold back the active glaciers. If they are not restrained and start advancing into the ocean and melting, this will contribute significantly to sea-level rise.

This, in fact, happened recently. In 2004 the ice floes that had fed the Larsen B Ice Shelf began moving faster toward the sea and started to thin. Since this occurrence, climate modelers, such as those at USGS and the NSIDC, are now putting this contingency into their prediction equations in order to increase the predictive accuracy of their models.

Scientists studying the glaciers on the ice sheet there today claim there is no longer a mass balance. Instead, there is a steady net loss of ice: More icebergs are calving into the ocean from rapidly moving ice streams than ice is being added to the continent via snowfall. While they are not ruling out the fact that the loss of ice mass is due to natural variations, they strongly support the theory that the main cause is anthropogenic (human caused).

Currently, the Ross Ice Shelf is only a few degrees too cool in the summer to experience a similar sequence of events, but scientists are keeping a close eye on it. Its dimensions are enormous. It spans an area 497 miles (800 km) across, encompassing 188,032 square miles (487,000 km^2). Larger in size than the state of California, it is monitored closely as seasonal temperatures in the Southern Hemisphere continue to climb.

Arctic

A similar scenario is taking place in the Arctic regions of the Northern Hemisphere. NASA is currently monitoring the polar ice sheets with the Ice, Cloud, and Land Elevation Satellite (ICESat), which was launched in January 2003. Three times a year this satellite uses a laser beam to measure the elevation of the ice sheets with a high degree of precision. NASA scientists have determined that 9 percent of the mass of Arctic sea ice is melting away each decade.

Meanwhile, a study conducted at the University of Colorado's NSIDC by Josefino Comiso found that most of the Arctic warmed significantly in the 1990s. It was also determined that the season when ice melt occurs (early spring to late fall) has gotten longer and warmer each decade. Climatologists at NSIDC view these trends as early warning signals of a changing global climate. Due to its responsive nature, the Arctic is one of the best places to detect the first serious indications of global warming, and as portrayed in National Geographic's film *Arctic*

Tale (2007), the Arctic is already undergoing serious changes to its eco-system, putting the wildlife in jeopardy.

In addition to global warming being the cause, researchers at NASA's GISS have also suggested that the increasing loss of Arctic sea ice may also be partly caused by changing atmospheric pressure and wind patterns over the Arctic that move ice around. This also serves to warm temperatures. It is believed that changes in air pressure and wind patterns may be caused by the high concentrations of greenhouse gases in the atmosphere, aggravated by the burning of fossil fuels.

Scientists at GISS point to several pieces of evidence that the Arctic is warming. The rate of warming since 1981 is eight times greater than the rate over the previous 100 years. The Arctic spring, summer, and autumn have all extended in duration, which has lengthened the prime sea ice melting period by 10 to 17 days per decade. A major problem with this is the positive feedback mechanism it causes. As temperatures climb, more ice is melted, which lowers the albedo, which, in turn, melts more ice. This feedback eventually changes the temperature of the ocean layers and impacts ocean circulation and salinity.

Whereas the melting of ice and consequent opening up of the Northwest Passage might open up new shipping lane access, which some may see as a benefit, it would be detrimental to wildlife habitat (polar bears, walruses, and seals), as it would impact their delicate eco-systems. It would also pose a hardship on the native communities that live in the area, because their culture, their lifestyle, and their survival would be placed at risk.

In addition, if Arctic permafrost is thawed, the soils could release vast amounts of stored carbon dioxide and methane into the atmosphere. If the oceans warm, the frozen natural gases currently stored on the seafloor may also be triggered to release because of the higher tempera-tures, which will also add more greenhouse gases to the atmosphere.

The Greenland Ice Sheet is another area of intense research today, visited by scientists from the United States, Canada, Britain, Austra-lia, Finland, Norway, Sweden, Russia, Denmark, and Iceland. Like the Arctic ice, this ice sheet begins to melt earlier in the spring than it ever had before. Usually, seasonal melt occurs along the edges at its lowest points. In 2002, the ice sheet melting accelerated. According to GISS, several scientists have been involved in research trying to determine whether the Greenland ice sheet is melting. Various scientists using different sources of data and different raw data processing techniques have generated a range of estimates depicting the mass balance of the Greenland Ice Sheet.

E. Hanna, a member of the Department of Geography at the University of Sheffield, in the United Kingdom; J. Box, of the Byrd Polar Research Center at Ohio State University; and P. Huybrechts, from the Department of Geography at the University of Brussels, Belgium, for example, used a combination of GRACE satellite data to calculate mass volume losses, airborne and satellite laser altimetry data analysis to calculate mass volume loss, and InSAR satellite radar interferometry to reveal widespread acceleration of glaciers. Jay Zwally, from NASA, used ERS radar *altimeter* data from airborne laser surveys to derive his measurements. The Climate Change Institute, supported by the NSF, NOAA, and NASA used a precision global positioning system (GPS) supplemented with ground survey and ice-core analysis and correlation. Although techniques and estimates varied among these studies, the end results all indicated that the Greenland Ice Sheet has lost hundreds of gigatons of mass in recent years.

These researchers agree that in the early 1990s, Greenland's ice sheet was nearly in balance. Since then, however, the ice sheet has become extremely out of balance, because it is currently losing a significant amount of ice to the ocean. The more they study global warming and how it relates to the equilibrium of the world's ice sheets, the more they try to develop sophisticated models to predict the future. They all acknowledge, too, however, that because the complicated physics of ice sheets are still not completely understood, the current models still are not able to portray all the complexity of nature.

An article appearing in the *New York Times* on January 8, 2008, titled "Melting Ice-Rising Seas? Easy. How Fast? Hard." expressed scientists's worry that Arctic ice melt may cause sea levels to rise more than two feet (0.6 m) this century—an estimate made also by the IPCC in its fourth report in 2007. The Arctic Council—an organization of countries within Arctic territory (Canada, Denmark/Greenland/Faeroe Islands, Finland, Iceland, Norway, Sweden, the Russian Federation, and the United States)—has commissioned a report on Greenland's situation to be completed prior to the climate treaty talks in Copenhagen, scheduled in December 2009.

A special three-day summit was held in Copenhagen beginning on February 10, 2009, by the Arctic Council in order to update the scientific community with the latest climate research information based on their report before the global political negotiations to be held in December (which are to be the formal successor to the Kyoto treaty).

Jonathan Bamber, an ice sheet expert at the University of Bristol, explained to the conference attendees that previous studies had misjudged the so-called Greenland tipping point (the point at which the ice sheet is certain to melt completely). "We're talking about the point at which it is 100 percent doomed. It seems quite an important number to get right. Such catastrophic melting would produce enough water to raise world sea levels by more than 20 feet (6 m)."

"We found that the threshold is about double what was previously published. It would take an average global temperature rise of 10°F (6°C) to push Greenland into irreversible melting. I'm not saying that if you have a temperature rise of 3.3°F (2°C) then you're not going to lose mass from Greenland, because you are. You warm the planet, ice melts." At the meeting in December, the world's nations plan to agree on a long-term solution for limiting human-caused global warming.

In August 2005, a 25.5 square mile (66 km²) piece of ice broke off the Ellesmere Island Ice Shelf in the Canadian Arctic near Greenland. This represents the biggest single ice loss in the past 25 years. The collapse even registered on earthquake monitors 155 miles (249 km) away. Scientists believe that the ice reaches some type of threshold, which triggers the fracture. What they wonder now is if the area has reached a critical threshold, and how much more ice may break away.

This ice shelf is only one of six in the Canadian Arctic, and global warming played a key role in its collapse. An expert in Arctic studies, Warwick Vincent of Laval University says that the remaining ice shelves in the area are currently 90 percent smaller than when they were first discovered in the early 1900s. Luke Copland of the University of Ottawa remarks that scientists are surprised at how fast ice shelves are reacting to climate change.

CRYOSPHERIC FACTS

Although most of Earth's ice exists in two remote polar locations, it affects the entire planet. It plays an important role in sea-level rise, storage of freshwater, and Earth's energy balance. The following list summarizes unique facts about Earth's ice:

- The melting of floating sea ice and the calving of glaciers into the ocean do not change sea level, because ice displaces about the same volume of water as it produces when it melts. The thinning and the retreat of glaciers on land do add water to the oceans.
- Antarctic ice is more than 13,780 feet (4,200 m) thick in some areas.
- The Antarctic Ice Sheet has existed for more than 40 million years.
- From 1950 to 2000, average temperatures in the Antarctic Peninsula increased by 4.2°F (2.5°C). This is equal to four times the global average.
- Over the past 100 years, almost 7,720 square miles (20,000 km²) of ice shelf was lost on the Antarctic Peninsula.
- Almost 90 percent of an iceberg lies below the water's surface—only 10 percent is above.
- Antarctic icebergs can calve icebergs more than 50 miles (80 km) long.
- The mass of ice is so heavy on Antarctica that the land underneath it sits 1.6 miles (2.5 km) below sea level.
- If all the ice on land melted, it would make sea levels rise more than 230 feet (70 m) worldwide.
- The mean height of the Greenland ice cap is 7,005 feet (2,135 m).
- Sea ice in the Arctic is getting thinner and covers less area by late summer.
- During the Ice Age, the sea level was about 400 feet (122 m) lower than it is today. Glaciers covered almost one-third of the land's surface.
- During the last warm period, 125,000 years ago, the seas were 18 feet (5.5 m) higher than they are today. Three million years ago, the seas may have been 165 feet (50.3 m) higher.
- If the ice on Antarctica melted, it would raise sea levels by 215 feet (66 m).

(continues)

(continued)

- If the ice on Greenland melted, it would raise sea levels 23 feet (7 m).
- The Greenland Ice Sheet covers 82 percent of the surface of Greenland.

As global warming continues and Earth's energy balance changes, the ice cover will continue to adjust, causing not only local but global effects on ecosystems, such as rising sea level, destruction of polar habitats, changes in the water cycle, and changes in surface albedo.

ICE CAPS

An ice cap is a dome-shaped ice mass that covers an area less than 19,305 square miles (50,000 km²) in size. Ice caps are very responsive to the effects of albedo. Normally, they act as "reflectors," redirecting the Sun's energy back out into space and keeping Earth cool. If the ice caps begin to melt, however, and darker surfaces are exposed, a negative feedback occurs, and Earth's surface begins to absorb the Sun's energy (heat) and accelerate the melting process.

The Arctic ice cap has been under great scrutiny the past several years. In 2007, for a time, the Northwest Passage shipping route over Canada and the Northern Sea Route over Russia was opened. During this time span, floating ice melted more than it had in more than 100 years. The results surprised scientists. No models predicted that the rate of melting and change would be as rapid as it has been. Scientists believe the response is tied directly to the rising concentration of greenhouse

(opposite page) This map depicts the extent of the melting of the Arctic polar ice cap from 1979 to 2007. Because of the extensive melting that has recently occurred, the Northwest Passage has been opened up to shipping traffic.

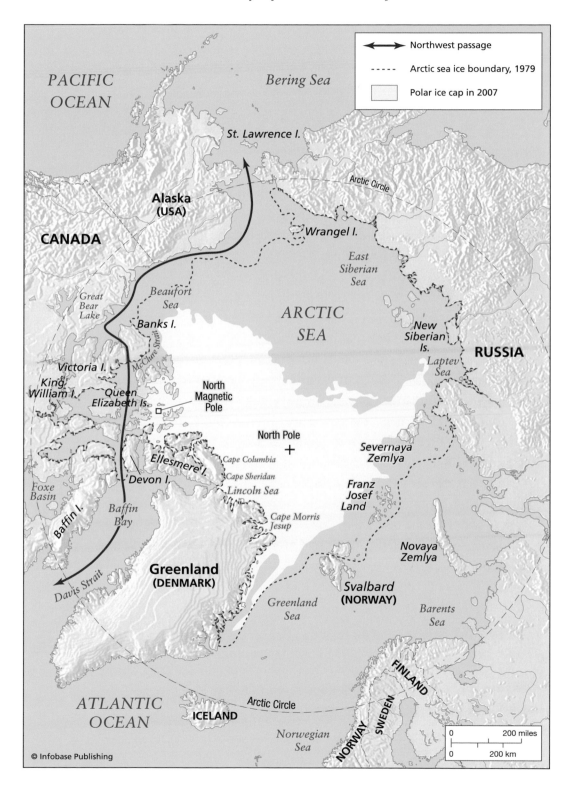

gases in the atmosphere. Some scientists believe the ice melt is completely accounted for by global warming. Others disagree, saying that while global warming is contributing a great deal, there could be other factors at play that scientists do not yet fully understand.

According to Jay Zwally, for example, a NASA climate scientist, "At this rate, the Arctic Ocean could be nearly ice free at the end of summer by 2012, much faster than previous predictions. The Arctic is often cited as the canary in the coal mine for climate warming. Now as a sign of climate warming, the canary has died. It is time to start getting out of the coal mines." He says it is the burning of coal, oil, and other fossil fuels that produces carbon dioxide and other greenhouse gases, responsible for human-made global warming.

According to Professor Wieslaw Maslowski, from the Department of Oceanography in the School of Engineering and Applied Sciences at Monterey, California, he believes that most models have seriously underestimated some key melting processes. He believes that models need to incorporate more realistic representations of the way warm water is moving into the Arctic Basin from the Pacific and Atlantic Oceans. According to Maslowski, "global climate models underestimate the amount of heat delivered to the sea ice by oceanic advection. The reason is that their low spatial resolution actually limits them from seeing important detailed factors. We use a high-resolution regional model for the Arctic Ocean and sea ice forced with realistic atmospheric data. This way, we get much more realistic forcing, from above by the atmosphere and from the bottom by the ocean."

Professor Peter Wadhams from Cambridge University used sonar data collected by Royal Navy submarines and says that the ice-albedo

Arctic sea ice has decreased about 14 percent since the 1970s. In this illustration, the extent of ice melt that has occurred on the Greenland Ice Cap from 1992 to 2005 covers an enormous area. By 2080, sea ice is expected to completely disappear during the summer months. The summer melting trend in Greenland has increased each year since 1979. Water flows through cracks to the base of the ice, lubricating it and causing it to flow toward the ocean. (*Modeled after Steffen/Huff, CIRES, University of Colorado, Boulder*)

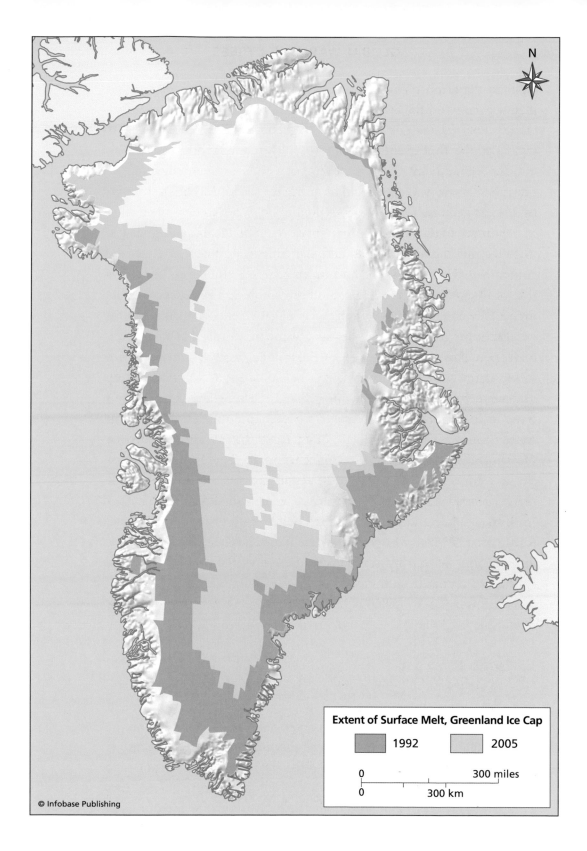

N

Extent of Surface Melt, Greenland Ice Cap

| | 1992 | | 2005 |

0   300 miles
0   300 km

© Infobase Publishing

feedback effect in which open water receives more solar radiation, which in turn leads to additional warming and further melting, is another process that needs to be considered. He believes this feedback effect is setting up the Arctic for further ice loss in the coming years. According to him, "The implication is that this is not a cycle, not just a fluctuation. The loss this year will precondition the ice for the same thing to happen again next year, only worse."

People see good and bad in recent developments of melting sea ice. With ice melting and opening up major shipping routes, it represents a boon for shipping, fishing, and oil exploration industries. The other side of the scenario is far from positive, however. The repeated melting and freezing cycles have extremely negative effects on the natural wildlife in the Arctic, such as the polar bear and walrus. It could destroy their fragile habitat and the delicate ecosystem.

A significant rise in carbon dioxide would trigger accelerated melting of the Arctic Ocean and Greenland ice caps, releasing large quantities of water into the oceans, in turn, triggering huge rises in sea level. It is projected that if the CO_2 level rose to 550 *parts per million (ppm)* by the year 2100, it could make the seas rise 8–37 inches (20–94 cm). If this were to happen, it would severely affect the coastal areas of continents and submerge areas such as New Orleans, in Louisiana, and Venice, Italy.

Through the analysis of satellite imagery and ground-based data, scientists at NASA have determined the Arctic has been steadily warming, causing the ice caps to shrink. In fact, over the past 30 years, the Arctic ice cap has both thinned and lost ice at its margins. NASA scientists expect the Arctic ice cap will shrink 40 percent by 2050 in most regions.

GLACIAL ISOSTASY

Glacial isostasy is the process by which Earth's lithosphere (crust and part of the upper mantle) is physically pressed down under the weight of an ice sheet. When the ice mass is later reduced or removed and the weight is lifted, it allows the crust to rebound (return to its original position). Similar to a person sitting on a sofa, when weight is applied to the cushions, the cushions depress to accommodate the weight. As soon as the weight (stress) is removed, the cushions return to their original equilibrium. Earth's isostatic process is the same when weighed down under ice (or rock); it just takes longer to complete.

When large masses of ice cover Earth's surface, it causes the crust to undergo both elastic and plastic deformation. Elastic deformation is temporary movement; the crust will return to its original position once the stress is removed. Plastic deformation is permanent deformity; it does not resume its original shape. The crustal rock gets displaced as the mantle sinks beneath it. An ice sheet 3,281 feet (1,000 m) thick has the ability to depress the ground beneath it 902 feet (275 m). It takes several thousand years for the isostatic adjustment to take place because there is a time lag between when the glacier builds up and Earth's crust responds. When an ice sheet begins to shrink, lifting the weight back off the depressed landmass, the depressed area then begins the process of rebounding, rising back up again. This process can take several thousand years, starting out quickly, then slowing down.

Glacial isostasy comes into play concerning global warming when calculating sea-level rise from the melting of ice sheets and the *thermal* expansion of water. In some instances, isostasy may offset some of the effects of sea-level rise, if adjustment is occurring. The problem, however, is that if isostasy is progressing slower than sea level is rising, offsetting any negative effects would be minimal.

There is evidence of past isostatic rebound. In the Baltic Sea and the Hudson Bay area near Canada, which were both covered with an ice sheet 14,000 years ago, ancient beach ridges exist today around 1,000 feet (300 m) above sea level. Scientists at NASA and USGS believe this area is still actively rebounding, just at a much slower rate than it did shortly after the last ice age. The Northern Baltic Sea is rising almost 0.4 inch (1 cm) a year, which equals about 40 inches (1 m) per century. For Hudson Bay to reach its state of "equilibrium," geologists calculate it still needs to rebound about 492 feet (150 m).

Currently, in Antarctica, the weight of its ice sheet is so immense that it has pressed the continent 0.6 mile (1 km) into Earth's crust. In Greenland, its ice is so heavy that the land beneath it has also been pressed down into a dish shape. If these ice masses were to melt, the land beneath them would also undergo isostatic rebound.

Ocean Currents and Climate

The circulation of Earth's oceans plays a significant role in Earth's climate. This chapter addresses the unique influence that the oceans play in shaping both the short- and long-term aspects of climate and why they have such an important effect on climate change. It examines major circulation patterns and the role of global heat transfer and distribution. It also looks at the unique cyclic circulation patterns that influence climate on a global basis. Without the ocean currents, Earth's climate would be very different, and understanding the behaviors of these currents is critical to understanding the effects of global warming.

UNDERSTANDING THE ROLE OF OCEANS IN REGULATING CLIMATE

The circulation of the atmosphere coupled with the ocean currents carry heat from the Tropics toward the poles. This may seem to be a simple, unalterable process, but there are several factors that can come into play to alter the heat-carrying circulation patterns and cause a significant change in the climate.

At one time, the importance of the world's oceans in shaping the climate was underestimated; but, in fact, the upper few feet of the ocean alone store as much heat energy as that stored in Earth's entire atmosphere. Due to its sheer size and complexity, however, study of the ocean was delayed until long after study of the atmosphere began. It was not until the 1800s that scientists first noticed the relationship between winds passing over the oceans and the moisture and warmth they brought to the adjacent landmasses.

This discovery spurred much scientific debate over the patterns of ocean circulation, namely, the method of circulation—differences in density owing to temperature or salinity or to persistent winds. According to the American Institute of Physics, thus began an era of discoveries and the birth of a new scientific field, oceanography. One American scientist, T. C. Chamberlin, made significant discoveries concerning the ocean's role in regulating the composition of the atmosphere. He determined that the warmer that ocean water became, the more carbon dioxide gas (CO_2) and water vapor (both greenhouse gases) it evaporated into the atmosphere. Conversely, the colder the ocean water, the more it absorbed both greenhouse gases. This was a major discovery, for the first time relating the role of oceans to glacial-interglacial episodes; as Earth warmed or cooled, the oceans contributed to the cycle by releasing or absorbing greenhouse gases. In their studies, Chamberlin and his students determined that this process was very complex, involving many intricate chains of chemical reactions.

As the burning of fossil fuels began to cause a more pronounced effect on the CO_2 concentration of the atmosphere, these theories challenged several assumptions in models of the 1900s. According to findings of the American Institute of Physics, many had assumed that nature could just take care of itself, and the more CO_2 added to the atmosphere, the more the oceans could absorb and accommodate. It was assumed that because the oceans held roughly 95 percent of Earth's CO_2, this "balance" was essentially limitless.

Simultaneously, it was also assumed that the general circulation of the oceans was so enormous that its flow would remain constant no matter what was done to the oceans or atmosphere. Oceanography as a science had a slow start because of its enormity and lack of funding to conduct research; there were still no government-sponsored research organizations in place to carry on this function.

Understanding the general circulation was a major achievement because of these technical and financial limitations. Initially, measurements of currents were obtained simply by throwing bottles into the oceans and monitoring where they traveled. It took a century just to determine the general circulation pattern. Initial discoveries were centered on the Gulf Stream because of its relatively rapid surface currents.

Through the 1950s, little funding was given to oceanographic research. Initial data gathering was centered on taking water samples from thousands of feet below the surface and calculating the temperature and salt content of the water. Through this work, it was discovered that ocean water sank near Iceland and then traveled along the ocean bottom to the South Pacific. By the mid-1950s, Henry Stommel, a renowned physical oceanographer, who later was affiliated with the Woods Hole Oceanographic Institution, discovered how cold, salty water sank in just a few northern regions and traveled along the ocean floor.

These discoveries of the 1950s proved to be significant: They marked the beginning of understanding what later became known as the general circulation model (GCM). This not only spurred interest but funding. Two important organizations were formed: the U.S. Navy's Office of Naval Research (ONR) and the International Geophysical Year (IGY). The ONR sponsored research on a variety of oceanographic topics. In 1957, the IGY began research on the role of the oceans in regard to climate change.

Through both organizations' research and achievements in the 1970s, information began to fall into place concerning short-term ocean-atmosphere feedback oscillations that operate on a timescale of a few years to a few decades, such as the El Niño Southern Oscillation (ENSO). It was finally possible to determine that El Niño events were connected to powerful climate events around the world, ranging from disastrous floods to crippling droughts. Once scientists such as

Wallace Broecker of the Lamont-Doherty Earth Observatory in New York understood these short-term oscillations, they began applying this knowledge to longer-term oscillations, such as glaciations. This began the era of climate modeling and prediction. Much of the data came from measuring carbon isotopes in samples of water collected worldwide. In building GCMs, they determined that ocean circulation took several hundred years to complete. Even more important, because this motion in the oceans was so slow, they were also able to determine its uptake was slow. It took hundreds of years for oceans to absorb the extra CO_2 that humans were rapidly adding to the atmosphere. The oceans could not keep up with the global warming rates of today.

In the mid-1960s, Peter Weyl, a scientist from Oregon State University, developed a unique theory on the triggering of ice ages. He determined that if the North Atlantic around Iceland were to suddenly become less salty—which could happen if global warming melted the Arctic glaciers and their melting freshwater emptied into the North Atlantic, diluting the upper ocean layers—it could effectively shut down the entire ocean circulation and trigger an ice age. The exact circulation pattern that enables the Gulf Stream to bring warmth northward from the equator to western Europe was eventually called the "thermohaline circulation pattern" or the "North Atlantic conveyor belt."

These initial discoveries eventually led to the development of several ocean-atmospheric computer models in the 1990s that illustrated how likely present global warming was to shut down the conveyor belt and plunge western Europe into an ice age. Since these studies, scientists have advanced further to determine that, in addition, Earth's tropical oceans could be just as important as the North Atlantic in the issue of rapid climate change. Instead of the issue being "regional" in nature, it is actually a "global" issue, where all regions are connected and changes in one area cause a ripple effect of response and change in other areas, driving a global feedback cycle. Broecker believed that these kinds of variations could cause major changes in the ocean-atmosphere system.

THE ROLE OF OCEANS IN CLIMATE CHANGE

The oceans and atmosphere are closely linked to each other and form the most dynamic part of the climate system. In the atmosphere, external *forcings* such as variations in the Sun's energy and concentration of greenhouse gases directly affect the circulation patterns and temperature of the ocean-atmosphere system. There is also an internal relationship. Because both the ocean and atmosphere are constantly in motion, they generate their own internal fluctuations. Short-term fluctuations in wind and temperature patterns directly influence the ocean's surface waters and are what cause local storm fronts. Fluctuations in the ocean have the ability to magnify, modify, or minimize atmospheric fluctuations. A small change in just one property of the ocean's characteristics (transportation, temperature, *upwelling,* currents) can result in major climate changes over large regions of Earth's surface.

According to Raymond Schmitt, a senior scientist in the Department of Physical Oceanography at the Woods Hole Oceanographic Institution, until the scientific world began to research and really understand the complexities of oceanography, the ocean was not given much weight in the study and prediction of climate. Initial models simply treated the oceans as a "shallow swamp," a source of moisture lacking any serious significance.

After years of extensive research, the oceans were viewed in their proper perspective and finally considered an equal partner to the atmosphere, with the realization that they both work in tandem to create the present climate by transporting heat from the equator to the polar regions. Oceans also have a much higher heat capacity than the atmosphere—roughly 1,100 times more.

The World Ocean Circulation Experiment (WOCE), a project supported by the NSF, has also provided overwhelming evidence that the oceans play a critical role in the climate. Based on the results of this experiment, it has been determined that the deep regions of the oceans have warmed significantly since the 1950s. In fact, roughly half of the increase in greenhouse warming predicted in models that has not been observed in the atmosphere has been absorbed by the world's oceans and held in this immense reservoir. Prior to this, it had not been considered that the oceans could store so much of the greenhouse gases on such short timescales.

Oceans have such a high heat-holding capacity that, according to Schmitt, it would take 240 years for the continual deposition of greenhouse gases to raise the ocean 1.7°F (1°C). It is because of this that

Schmitt stresses that the oceans be recognized for the critical role they play in global warming. Their involvement goes well beyond just delaying the process by sequestering CO_2. Research done at the Woods Hole Oceanographic Institution has also made clear how the slow movement of cold and warm regions of ocean water can play a key role in weather on a cyclic basis for months at a time. A prime example of this is El Niño.

Because oceans are dynamic, they interact with the atmosphere in two distinct ways—physically and chemically. The physical component occurs through the exchange of heat, water, and momentum. Because the oceans cover more than 70 percent of Earth's surface, their effect is significant; they act as the storage bins of huge amounts of energy—heat energy. Unlike land, which heats up and cools rapidly, water changes temperature much more slowly. It takes much longer for a body of water to heat, but once it is heated, it will stay heated for a much longer time than land will, even if the source of heat is removed. This is called a large temperature inertia. Because of this inertia, oceans play an overwhelming part in the balance of Earth's heat energy. Earth's oceans are its global heat engine.

In this dynamic ocean-atmosphere system, Earth's oceans heat up over time and store heat, which eventually escapes into the atmosphere, warming it. As the air above the ocean warms in relation to adjacent colder air over land, it creates a temperature gradient, which in turn causes wind. When wind blows across the surface of the ocean, it moves the surface and begins moving the ocean in a horizontal current pattern. But heat is not the only force at work on the water. Temperature and salinity are also working on the vertical column, or depth, of the water. This causes vertical current patterns because warmer, less salty water flows upward and colder, saltier (denser) water sinks.

In deep ocean waters, the density of the water, which is directly controlled by temperature and salinity, controls circulation patterns. The vertical circulation cells that form as a result allow heat to be stored in the great depths of the ocean where they can be later released back into the atmosphere.

With all these vertical and horizontal processes working together, a complex circulation is formed that causes the warm surface waters to travel toward the polar regions. These currents on the surface release heat to the atmosphere during their journey, effectively heating the nearby landmasses. In order to offset these warm surface currents, cold deep currents form and travel from the poles back toward the equator to repeat the cycle.

According to scientists at NASA, the oceans and atmosphere work together to distribute heat and regulate climate. Because the oceans have such immense thermal capacity, it allows them to slow the rate of climate change. The upper 10 feet (3 m) of ocean water hold as much heat as the entire atmosphere. This is what makes some coastal areas warmer, even though they may be located fairly far from the equator. A prime example is western Europe. While it is quite far north, it has a much more mild climate because of the warmth from the passage of the Gulf Stream. Without this warm northward current delivering equatorial heat on its way toward the North Pole, Europe would be much colder. As a comparison, the Scandinavian countries are at the same latitude as Alaska but have a milder climate.

The world's oceans also store and transport CO_2. The oceans have absorbed about half of the total CO_2 added to the atmosphere during the last 100 years by human activities such as the burning of fossil fuels and deforestation. This "sequestering" of carbon is a slow process, however, and will not keep up with current rates of CO_2 input into the atmosphere. Phytoplankton in the ocean also stores CO_2 from the upper layers of the ocean in their carbonate shells. Eventually, this CO_2 settles to the ocean floor and gets buried in the sediment there.

According to NOAA, heat and CO_2 are also exchanged vertically in the ocean from the surface to the deep bottom layers through a process called "upwelling." This vertical circulation occurs in coastal areas and is the mechanism that makes many coastal areas very productive in fishing resources. The longer-term trends in upwelling may be related to global warming. The upwelling process moves the cooler nutrient-rich water up toward the surface, where it replaces the original surface water. This is the mechanism that causes the west coasts of all continents to have relatively cool surface waters. It also provides fertilization to these waters, increasing their biological productivity.

Climate modelers at NASA have done a considerable amount of work relating the ocean's physical properties to global warming. They have determined that during the 1900s, the ocean played a significant role in regulating the atmospheric temperature due to global warming. In fact, if the oceans were not so efficient at absorbing CO_2 and heat, the temperature rise experienced in the past 100 years on land would have been doubled due to increased greenhouse gases. NASA scientists have concluded that the ocean's tremendous storage capacities have offset— so far—some of the negative effects of anthropogenic global warming. What they do not know, NASA scientists stress, is whether the ocean's role as "climate moderator" will persist over the long term. There is evidence that Earth's climate has reached strategic tipping points in the past and has not been able to respond fast enough, triggering a glacial period or a drought. One thing NASA scientists caution is that as the

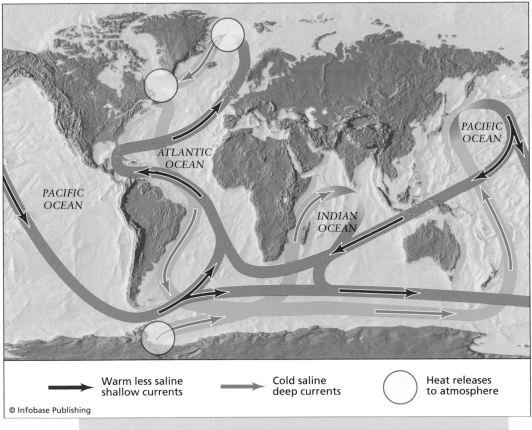

| → Warm less saline shallow currents | → Cold saline deep currents | ◯ Heat releases to atmosphere |

© Infobase Publishing

Working as a massive conveyor belt of heat, the oceans' thermohaline circulation has a significant effect on weather worldwide.

rate of freshwater increases near the North Pole, it will raise the freezing point. More surface ice would keep heat from being released to the atmosphere. If this occurred, it would make it more likely that the North Atlantic conveyor belt would be slowed or halted, triggering a climate change disaster. (See chapter 6 for a more detailed discussion on abrupt climate change.)

The second way the ocean and atmosphere interact is chemically. When water evaporates from the ocean's surface, clouds are formed. Water vapor has a twofold effect: 1) water vapor is a greenhouse gas, so it plays a role in heating the atmosphere, but 2) it also forms clouds, which block incoming solar radiation, thereby cooling Earth. Over an extended period of time, it is not known if the net effect from water vapor on global temperatures will be cooling or heating.

Even more important is the ocean's role with CO_2. According to NASA, most of the world's carbon is located in the ocean. Because of this, the exchanges that happen between the upper and lower levels of the ocean, as well as the ocean surface and the atmosphere, are very important. Natural chemistry processes play a large part in what happens to some of this carbon, but biological processes also play a factor, and they are important to climate change. The process of *photosynthesis* turns CO_2 into organic material. When it is in the ocean, it sinks to the ocean floor in a process that scientists call the biological pump. The way it works right now is as a *carbon sink:* CO_2 entering the ocean is stored on the bottom. Scientists have proposed that if the ocean's circulation patterns were disrupted, this carbon could be released back into the atmosphere, making the oceans a CO_2 source instead of a sink.

One of the main issues climate modelers are working on today is how the physical and biological processes of the ocean will respond to chemical and physical changes in the atmosphere. According to experts at NASA, this raises the following questions:

1. Will an increase in storms cause the upper ocean waters to mix more?
2. How will the phytoplankton react? Will they be more or less productive?

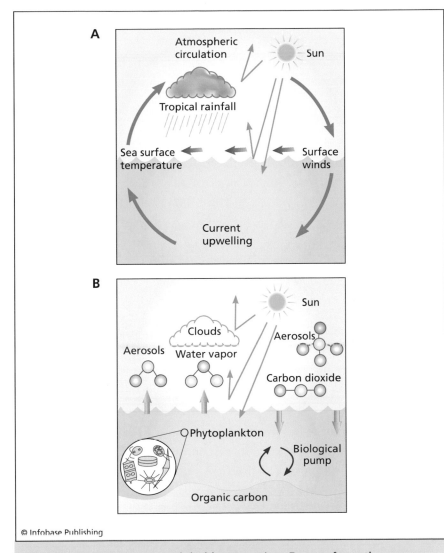

© Infobase Publishing

Top: The ocean acts as a global heat engine. Energy from the ocean forms heat and water vapor. As the atmosphere warms, pressure gradients form surface winds that drive ocean currents. *Bottom:* The oceans and atmosphere interact chemically, enabling the ocean to absorb more than 90 percent of the world's carbon. This transfer process is referred to as a biological pump.

3. Will sediments high in iron content blown into the ocean by wind act as "fertilizer" to the phytoplankton and cause a rapid growth?

4. Will climate change increase the dust added to the ocean?
5. Will these changes increase activity of the biological pump, which will increase the removal of CO_2 from the atmosphere by the ocean?

One theory, indicated by research, is that if phytoplankton increased in the ocean, it would counteract the increase of CO_2 in the atmosphere, slowing it down and offsetting global warming. One area that researchers agree needs a lot more research in order to answer critical questions concerns the marine food chain and its vulnerability to short- and long-term climate change. Scientists at NASA are currently working to find answers to these questions.

CYCLIC OCEAN CIRCULATION THAT INFLUENCES CLIMATE

There are cyclic circulation phenomena that occur in the ocean that play a direct role in climate over large areas of Earth. The most significant are the North Atlantic Oscillation (NAO) and ENSO. The most powerful of these cyclic anomalies is NAO. While it does not produce the violent weather effects as dramatic as those of El Niño, NAO is much more consistent. NAO is a system involving two separate pressure systems and their relationship. The first is a high-pressure system that sits over the Azores, a group of islands 900 miles (1,448 km) west of Portugal; the second is a low-pressure system over Iceland. Both of these pressure systems generally exist on a year-round basis. During the spring, summer, and fall, the two systems generally stay weak and do not interact. During the winter, however, they both come to life, and it is their direct interaction that controls the climate over the Atlantic Ocean and all the surrounding continents.

According to NOAA, during the winter months, both pressure systems fluctuate in pressure relative to each other. The greatest effect of NAO is on the storms passing into Europe. Between *cyclone* circulation patterns over the Iceland and Azores pressure cells, there is an area where they come together to form a steady, strong, forward-moving current of air that channels weather systems from the United States directly into Europe. NAO has the most dramatic effect on storms from December through March each year.

When the pressure difference between the two systems is large, northern Europe has higher temperatures, the Middle East has droughts, and the northeastern United States has warm temperatures. Rainfall in Europe can increase by 0.14 inch (0.36 cm) per day, and temperatures can rise by 5°F (3°C). If the pattern persists, it can extend the growing season up to 20 days in Sweden. When the pressure difference between the two systems is small, the Mediterranean is rainy, Scandinavia is extremely cold, and the East Coast of the United States is cooler.

According to scientists at NASA, NAO varies in a rhythmic pattern from decade to decade. They have determined that since the 1960s, the difference in pressure has noticeably increased for three to five years, and then has noticeably decreased for three to five years, setting up a recognizable cycle. Ocean currents or the formation of sea ice have been suggested as possibilities for this cyclic pattern that has prevailed now for almost half a century. Scientists at NASA are currently working on models to better understand NAO, which would allow them to predict its behavior and allow affected countries time to plan ahead; for example, if a year for poor agricultural productivity seems probable, more food could be stored in reserve for those lean years.

Based on computer modeling work done so far, scientists at NASA believe that changes in NAO are directly related to sea surface temperatures. They are currently working on models to integrate the effects of ocean currents and sea ice, as well as trying to determine whether global warming is playing a part.

El Niño is an abnormal warming of surface ocean waters in the eastern tropical Pacific. It is one part of a larger system called the Southern Oscillation. The Southern Oscillation is a back-and-forth pattern of reversing air pressure that occurs between the eastern and western tropical Pacific and affects the weather worldwide.

When the surface pressure is high over the eastern tropical Pacific, it is low in the western tropical Pacific; conversely, when it is high over the western tropical Pacific, it is low over the eastern side. The phenomenon works like a switch, with the pressure system alternating back and forth. This is where it gets the name *oscillation*.

Usually, winds blow from the east to the west along the equatorial region in the Pacific. Because the wind pushes the water, it actually piles up the water about 1.7 feet (0.5 m) in the western portion of the Pacific. As a result, the ocean on the east (next to South America) has the deeper, colder water pulled up from below to replace the water that was pushed westward (toward Indonesia). This means that in normal years, the western Pacific has warmer waters and the eastern Pacific has colder waters.

During an El Niño, however, the winds that push this mass of water weaken, which causes some of the warm western water to flow back toward the east, which, in turn, means that less cold water is brought up from the depths. As a result, part of the buildup of water in the west flows back to the east. Because of this and the fact that the cold upwelling is minimized, the water in the eastern Pacific becomes warmer, a major component of ENSO.

Once this process occurs, the winds pushing the water get weaker, causing some of the water pushed to the west to flow back to the east. Once the ocean water's temperature rises, it causes the winds to significantly weaken, making the ocean even warmer. This sets up a positive feedback cycle, perpetuating an El Niño pattern. When ENSOs stay strong, the winters in the southern United States are usually very wet, and Indonesia experiences drought conditions. Australia has experienced increased rainfall and flooding.

ENSOs can persist for a year or more and historically have occurred on a three- to seven-year cycle. Computer modeling algorithms today are able to predict an ENSO up to a year and a half in advance. In order to predict occurrences, NOAA operates a network of buoys that measure currents, temperature, and winds. The data collected by the buoys is transmitted to NOAA research facilities in real time.

ENSOs can be detected through the analysis of sea surface temperatures and other collected data. During normal conditions, the western Pacific Ocean is warm, and the eastern Pacific Ocean is cool, with a cold tongue on its extreme eastern edge. The winds in the western Pacific are very weak, and the winds of the eastern Pacific are blowing toward the west. During El Niño, warm water spreads from the western Pacific Ocean (Indonesia) toward the east (South America). The normal cold tongue has weakened, and the winds in the western Pacific, which are usually weak, are blowing strongly toward the east, pushing warm water with them. This makes the water in the center of the Pacific Ocean warmer than normal. Sometimes an event called La Niña follows El Niño. La Niña is a cold event; the cold tongue is cooler than usual by around 5°F (3°C).

Although El Niño has been studied extensively, the exact mechanisms that control it are still not completely understood. According to NASA, not all ENSOs are identical, and the atmosphere does not always react in the same way. Therefore, each episode's characteristics are not predictable, which is why El Niño seems to get blamed for so many unexpected, sometimes violent weather events. Climatologists are becoming better at determining when El Niño events are going to occur, about a year ahead of time.

Rising Sea Levels

Of all the potential results of global warming impacting humans, a rise in sea level is viewed as one of the most serious, harmful, and destructive. Rising sea level has the potential to have a negative impact on human survival, environmental health, and economics, hence the lifestyle of millions of people worldwide. Because of this, understanding sea-level rise—through measurement, observation, and interpretation—has been a key focus of many climatologists during recent years. Sea-level rise accelerated during the 1900s. The increase of greenhouse gases in Earth's atmosphere from anthropogenic sources has warmed the environment enough that the global average sea level has risen 4–8 inches (10–20 cm).

As the atmosphere has warmed in recent decades and glaciers and ice caps have melted, it has directly affected ocean levels. It is expected that sea level rise will continue, and even accelerate, if global warming continues at the same pace. At one time, the large polar ice sheets were assumed to be stable, but that is no longer the case. Current models are being refined as Earth's masses of ice react to changing temperatures. Predicting future sea-level rise is difficult as climatologists work to understand the dynamic behavior and delicate balance of Earth's enormous, changing ice sheets.

This chapter looks at the issues surrounding sea-level rise. It outlines what sea-level rise is and how it is calculated, then presents a historical perspective. Next, it addresses variations in sea level and the effects change will have in different areas of the world. It concludes by touching on how experts are attempting to make future projections.

SEA-LEVEL RISE

Sea levels are currently rising and are expected to continue rising, possibly even at an accelerated rate, over the next century and beyond. It is not caused by one simple mechanism; it is the result of several processes, such as:

- the melting of glaciers and ice caps from continents (the melting of ice already in the water does not affect the sea level)
- thermal expansion of the oceans' upper layers
- the melting of the Antarctic and Greenland Ice Sheets
- redistribution of terrestrial water storage
- oceanographic factors, such as changes in ocean circulation or atmospheric pressure
- vertical land movement

The Intergovernmental Panel on Climate Change (IPCC), in its fourth report, issued in 2007, predicts that global mean sea level will continue to rise. The exact amount will not only depend on the above factors but will also be affected by the anthropogenic factor, specifically the emission of fossil fuels. If anthropogenic factors are not kept in check, melting and thermal expansion will increase, causing sea levels to rise further. The table on page 94 reflects the IPCC's projections of future sea-level rise.

The two main processes that contribute to sea-level rise are thermal expansion and the melting of glaciers, ice caps, and ice sheets. The more

Sea-Level Change (inches/centimeters)			
YEAR	LOW	MEDIUM	HIGH
2050	4/10	6/16	9/22
2100	8/20	14/35	20/50
Source: IPCC, 2007			

heat water absorbs, the more volume it occupies; the density of seawater is determined by temperature. Thermal water expansion is one of the few aspects associated with global warming that can be calculated directly from basic physics. When ice melts, it is added to the ocean through surface melting and runoff, as well as ice directly flowing into the ocean.

According to the IPCC, one of the biggest uncertainties of future sea-level rise is how the large ice sheets of Antarctica and Greenland will behave as the atmosphere continues to warm. There is enough ice on Greenland and the western portion of Antarctica to raise sea levels by 40 feet (12 m) should it melt. Even if this amount of ice melted over a long period of time, it would impact millions of people who live along the world's coastlines. According to James E. Hansen of GISS, these ice sheets' potential collapse could be the biggest long-term risk that humans face from global warming. More than 100 million people live within 3.3 feet (1 m) of mean sea level. In fact, some island states exist entirely at that low elevation, and their existence is threatened by sea-level rise. Coastal wetlands are also endangered. Property destruction is a serious issue, and many countries that would be negatively impacted could not afford the economic impacts of the destruction caused by rising water levels.

If there were to be substantial melting of the Greenland or Antarctic ice cap, sea levels would increase for centuries. Examining proxy data in rock formations, geologists at USGS have identified ancient beaches that existed far above the present sea level, confirming these areas were formed during warmer climates of Earth's past. During the last major interglacial period 125,000 years ago, Earth's temperature was comparable to Earth's predicted temperature in the next few centuries due to global warming. Sea level 125,000 years ago was 20 feet (6 m) higher than it is today (during this time, Antarctica was still covered in ice). It has been estimated that if the ice on Greenland were to melt, present sea level would rise roughly 20 feet (6 m), destroying the world's coastal communities.

Another issue the IPCC brought up in its 2007 report concerned the concept of ice surges. Analysis of satellite radar data in 2006 determined that the velocities of large ice streams in southern Greenland had doubled over the past five years and then slowed again. These sporadic surges surprised many of the scientists studying them, leading them to believe that the surges were more sensitive to global warming than initially thought. In addition, gravitational force measurements collected by the satellite determined there were significant changes in the mass of the ice sheets on a yearly basis, indicating that both ice sheets were losing significant amounts of ice to the ocean. In its 2007 report, the IPCC for the first time acknowledged the serious possibility of ice surges and their role in potentially adding significantly to sea-level rise before the year 2100, as well as beyond.

MEASURING SEA LEVEL

The earliest gauges used to measure the tides were simply measuring sticks attached to piers. By the mid-1800s, tide gauges using floats began to be used. Much of the difficulty in gathering data to calculate sea level was centered on the fact that the geographic distribution was weak; there were simply not enough gauges distributed around the world to enable a reliable global model to be built. Most of the reliable records came from the United States and Europe. In addition to widespread geographic distribution, time interval is also an important factor. The longer the time interval that records are collected at gauging sites, the better, because long-term observations can be used to infer more accurately significant trends.

Tide gauge records depict not only the general global trend but also geographical and temporal variations specific to the gauge's location. These local fluctuations can be the result of *interdecadal* fluctuations of ocean density and circulation, isostatic adjustment of the land from the previous ice age (when the weight of the ice is removed from the land, the land relaxes again and rises up because the pressure weighing it down has been removed), or *subsidence,* which can happen when underground fluids (groundwater, oil, and so on) have been extracted, making the land settle. According to NOAA, an area that these characteristics have successfully been applied to is in the middle Atlantic region of the U.S. east coast.

Nowadays, sea-level measurements can be obtained globally from satellite data. In 1992, through the joint effort of NASA and France's National Center for Space Research (CNES), *TOPEX/Poseidon* was launched and successfully collected data until October 2005. On December 7, 2001, the *Jason-1* satellite was launched by NASA, continuing the data-gathering mission started by the *TOPEX/Poseidon.* The satellite space program at NASA for gathering data to calculate global sea level is called Ocean Surface Topography. According to Jorge Vazquez at NASA's Jet Propulsion Laboratory, satellite oceanography and altimetry is a new, evolving science that has revolutionized and changed the way scientists obtain global sea levels.

An altimeter is used to collect the data. It is a microwave radar pulse that is transmitted from an orbiting satellite in space. When the pulse hits its target on Earth, the signal is bounced back to the radar. The round-trip travel time of the radar pulse is recorded. This is similar to the radars that the highway patrol uses to catch speeding cars. It is also the methodology that Doppler radar uses to detect where rainstorm activity is currently occurring. The further away an object is, the longer it takes for the return pulse to reach the sensor and be recorded.

According to Vazquez, the use of this concept for the oceanographic sciences has revolutionized the way scientific knowledge is collected today. The function of the altimeter is to measure the height of the ocean. Although the ocean may seem level in its immense, vast state, it is far from it. Because there are physical structures, such as seamounts and trenches that influence the gravitational pull around them, it affects sea level. The Moon and the Sun also affect the ocean's surface level by pulling on it. Other factors also come into play, such as winds, ocean currents, thermal expansion from the Sun's heat, and atmospheric storms. Each of these factors changes sea level by a specific amount.

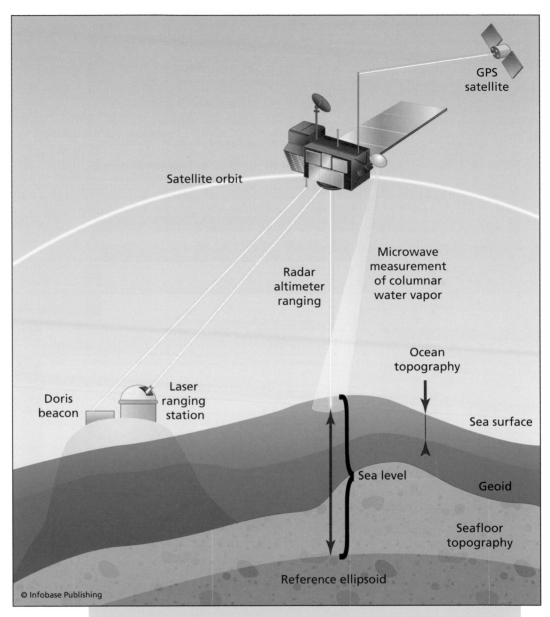

The *TOPEX/Poseidon* satellite sent out a radar signal to the ocean below, received the signal back, determined the height, and transmitted the reading to a real-time GPS.

The *TOPEX/Poseidon* was considered a technological breakthrough for NASA, because for the first time, it allowed scientists to determine by what amounts all of these factors influenced sea level. It was also from this data that scientists have been able to gain much of their understanding of the El Niño phenomenon. This satellite enabled scientists at NASA to observe how the ocean changes from one year to the next and study climate change and global warming. In 1998, a global map was built denoting sea-level change in the entire Pacific based on data from the satellite. *TOPEX/Poseidon* was able to detect changes of only one inch (2 cm), which was a remarkable improvement over a previous satellite, *Seasat,* which was used in 1978 and could only detect changes on the order of several yards. The *TOPEX/Poseidon's* accuracy was attributed to the quality of its laser beams in combination with a global positioning system (GPS).

Oceanographers and climatologists at NOAA are now able to study and determine the relationships between sea-level rise and ocean sur-

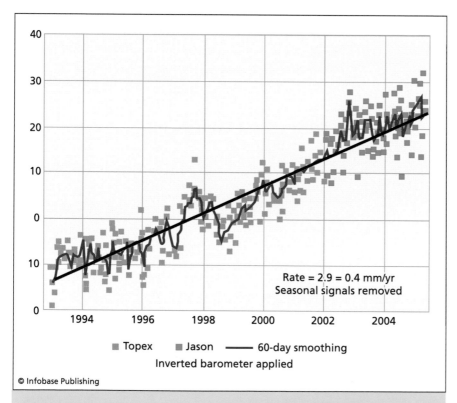

Rate = 2.9 = 0.4 mm/yr
Seasonal signals removed

■ Topex   ■ Jason   —— 60-day smoothing
Inverted barometer applied

© Infobase Publishing

Global mean sea-level rise as measured from the *TOPEX/Poseidon* and *Jason-1* satellite. These systems allow scientists to study climate change due to global warming. *(Modeled after University of Colorado)*

face temperature. This allows researchers to better understand phenomenon such as ENSO and the Pacific Decadal Oscillation (PDO). Scientists, both at NOAA and NASA, are currently looking at detecting changes in sea level due to changes in global climate, specifically on how the average global sea level rises during a given year.

Maps have been prepared using data from both *TOPEX/Poseidon* and *Jason-1*. Both NOAA and NASA scientists have detected complex patterns of sea-level rise using these satellites. Besides seeing patterns supporting the PDO, they have also detected a pattern in the North Atlantic caused by a slowdown in the circulation of the subpolar gyre (a clockwise circular ocean current). This has experts at NASA concerned because it can lead to a decrease of the northward heat transport of the ocean in the great conveyor belt. In addition, they have also detected a trend in the Indian Ocean of a decrease in the northward flow of the upper ocean, which is linked to a long-term warming of the upper Indian Ocean. Data from these satellites supply scientists at NOAA and NASA with information on not only sea surface height (topography) but the transport of heat, ocean circulation, wave height, ocean tides, wind speeds, and climatic events. This data can then be used in models designed to detect global climate change.

A HISTORICAL PERSPECTIVE

During the last ice age, huge ice sheets covered parts of North America, northern Europe, and parts of Asia. Because so much of Earth's water supply was locked up in ice, global sea level was 394 feet (120 m) lower than it presently is. Once the ice sheets began to melt, sea level began to rise. During that process, there were intermittent intervals when sea levels rose in rapid spurts called "meltwater pulses." One such pulse occurred between 14,600 and 13,500 years ago, increasing sea levels by 52–79 feet (16–24 m). Water originated from both North America and Antarctica. Rates of melting and sea level rise then declined after that during the Younger Dryas cold period, followed by another surge 11,500 to 11,000 years ago, when sea level may have risen as much as 92 feet (28 m). Another meltwater pulse occurred 8,200 to 7,600 years ago. By the mid-Holocene Period (6,000–5,000 years ago), glacial ice age melting had run its course: In the following years, there was very little sea-level rise—until recently.

According to NASA, 20th-century sea-level trends are significantly higher than the trends of the past few thousand years. Based on the analysis of coastal sediments retrieved from several locations, a new phase of sea-level rise began in the mid to late 1800s and has continued. Based on readings taken from tidal gauges in coastal harbors, global sea level has been increasing about 0.07 inch (0.17–0.18 cm) per year, which scientists attribute to global warming, specifically the melting of the world's glaciers and the thermal expansion of water. Recent analysis of the *TOPEX/Poseidon* data indicates a trend of a 0.1-inch (0.3-cm) increase per year in global sea-level rise.

According to NASA, an area of concern based on recent satellite observations of the Greenland and West Antarctic Ice Sheets is a thinning at the lower elevations, which is causing glaciers to empty at a faster rate into the ocean. Researchers have determined that this alone has added 0.08–0.02 inch (0.02–0.06 cm) per year to the oceans within the last decade. If either ice sheet melted completely, it could raise sea level by 16–23 feet (5–7 m). If a global temperature increase of 3.3–8.3°F (2–5°C) occurred, it could destabilize Greenland enough that it would cause an irreversible melting and trigger that magnitude of a sea-level rise. This magnitude of temperature rise falls within those predicted in climate projections for the 21st century. NASA clarifies, however, that any significant meltdown would take many centuries and views it as highly unlikely that annual rates of sea-level rise would exceed those of previous postglacial meltwater pulses.

VARIATIONS IN SEA LEVEL

There are several factors that can cause variations in sea-level rise, such as the following:

- the rebound (or rising up) of Earth's crust after the melting of ice from the last ice age
- plate tectonics and volcanism raising the height of the land's surface
- local subsidence of Earth's crust from groundwater extraction
- ground subsidence from sediment compaction
- changes in atmospheric wind patterns and ocean currents

One factor that must be taken into account when calculating sea-level rise is the mass balance. This is the difference between ice input and output in the glacial-ocean system. In other words, the amount of snowfall received in Antarctica and Greenland must be taken into account and calculated against the amount of ice that is melting into the ocean. If there is more melting of ice sheets than there is accumulation of snow on ice caps, then sea levels will rise. If the balance is the opposite, sea level will drop. Just as when all the ice cubes in a glass of lemonade melt, the level of the drink does not rise up and spill out of the glass, floating ice does not affect sea level when it melts, because it was already displacing the water at that location. Because of this, melting of the northern polar ice cap—which is floating pack ice—would not make sea levels rise. But just because it will not affect sea level does not mean melting the ice will not matter. Other, related melting ice in the area will affect sea level.

Although there are several mechanisms that cause short-term changes in sea level on the order of hours to weeks to months—such as storm surges, changes in water density and currents, El Niño events, and seasonal variations—it is the longer-term changes that affect global sea level that are receiving attention as they pertain to global warming. Along with water locked up in snow and ice, thereby affecting global sea level, geologic influence is another long-term factor that influences sea level. During an ice age, when the bulk of water on Earth is sequestered in ice, sea levels are much lower. This has also held true when Earth's continents have been nearer to the poles during past plate-tectonic periods. When the landmasses were tectonically situated nearer to the equator, and the majority of water was not frozen, sea levels were much higher than they are today. Over the past several million years during glacial-interglacial periods, sea level has varied by hundreds of feet. Long-term changes that climatologists must keep in mind when studying global warming and assessing sea level changes are the following:

- Isostasy, or changes in the uplift or subsidence of Earth's crust after or during glaciations. As ice melts and the land's surface is relieved of the tremendous weight of the ice, the land will slowly rise as it returns to its previous elevation before the weight and stress of the ice was added. This can offset sea-level rise because the land is also rising in elevation.

- Change in the quantity of ocean water. Changes in sea level occur as precipitation (usually in the form of snow) is added to the Greenland and Antarctic Ice Sheets. As the temperature of the atmosphere increases, additional melting of ice caps and glaciers will occur, affecting the mass balance, adding additional water to the ocean, and causing sea levels to rise.
- Change in the configuration of the ocean basins. As plate tectonics carries on its continuous movements, changing the shape and elevation of the ocean floor and sides, this affects the shape of the basin, which in turn affects sea level. Tectonic uplift can also play a part and change sea levels.

According to an article that appeared in the *New York Times* on January 8, 2008, Greenland's ice is melting enough each year to become unstable and is now a concern to scientists for its role in potential sea-level rise. During the spring and summer months, the meltwater lakes and rivulets on the ice sheet absorb up to four times more energy from the Sun as unmelted snow. They form natural drainpipes, called "moulins," that penetrate straight through to the bedrock below. This process serves to lubricate the sheet of ice and helps move it faster toward the ocean. As a result, huge chunks of glaciers have been breaking off into the ocean, especially along the west coast.

According to Ted Scambos of the National Snow and Ice Data Center in Boulder, Colorado, scientists are concerned that if this process speeds as global warming increases, that sea levels will continue to rise, negatively impacting coastal areas worldwide. But, not all climate scientists share the viewpoint that the ice is unstable enough at this point to be of immediate concern. Richard Alley, from Pennsylvania State University, believes that the Greenland Ice Sheet has fluctuated throughout history and its current reaction to warming is simply part of a natural cycle.

Most important, however, is that no matter what experts' short-term views are, they seem to agree on the long term: If current greenhouse-gas *emissions* continue at the level they are today, the resulting global warming and loss of ice at Earth's polar regions will negatively impact the world's coastlines for centuries. Because of this, Eric Rignot of NASA's Jet Propulsion Laboratory stresses the importance of both government policymakers and the general public becoming aware of the issues and cutting back emissions from fossil fuels before it is too late. According to Rignot, if global warming is not slowed, sea levels could rise three feet (1 m) just from water flowing off Greenland, another three feet (1 m) from Antarctica, and 18 inches (0.5 m) from alpine glaciers. As Rignot says, "Things are definitely far more serious than anyone would have thought five years ago."

According to a report in *USA Today,* scientists are focusing on the West Antarctic Ice Sheet because water melted off of it could force global sea levels to rise an extra 20 feet (6 m). Because the ice sheet's bottom is mostly below sea level, the West Antarctic sheet is considered more likely to collapse than the East Antarctic sheet. The west sheet's ice moves onto the Ross and Ronne Ice Shelves, which are floating on the ocean. The Ross Ice Shelf is huge; it measures 450 by 600 miles (724 by 966 km) and up to 4,000 feet (1,219 m) in thickness. If global warming continues to warm the atmosphere and oceans, it could melt the shelves.

Scientists monitor the flow of ice from the West Antarctic sheet to the shelves as they travel in channels called ice streams. According to Robert Bindschadler of NASA's Goddard Space Flight Center in Greenbelt, Maryland, it is important to monitor the ice streams to get an idea of where the most unstable areas are. The scientists there place GPS receivers on the ice rivers in order to receive location and elevation data from the satellite, precisely describing ice movement.

Currently, scientific consensus, based on the Goddard study, is that the ice at Antarctica is close to being "in balance" and not a threat at this point to sea-level rise. According to Michael Oppenheimer of the Environmental Defense Fund, he believes it would take several hundred years for Antarctica to melt, if global warming were to reach that point.

Scientists are concerned, however, that global warming could make the Ross Ice Shelf become thinner over time. A worst-case scenario would be for the shelf to disappear, leaving the Antarctic Ice Sheet vulnerable to collapse and melting of ice, which could hypothetically raise sea levels up to 20 feet (6 m) in 250 to 400 years. Oppenheimer believes the most likely scenario is that the Ross Ice Shelf will gradually melt over the next century and be gone within the next 200 years. During this melting period, the Antarctic will add up to seven inches (18 cm) per century to global sea-level rise.

Because of the important role that the Antarctic Ice Sheet plays in global warming and potential sea-level rise and the fact that the mass balance (whether it is shrinking or growing) is not well understood, understanding the ice's dynamics has been given a very high priority by the Polar Research Board of the National Research Council, the Scientific Committee on Antarctic Research (SCAR), and the NSF's Office of Polar Programs. The illustration on page 105 shows the various ice shelves currently being studied.

While the ice sheets represent a large potential source of sea-level rise, another source is from the long-term melting of the world's glaciers. According to the USGS and IPCC, the following tables depict the estimated potential maximum sea-level rise from the total melting of today's ice, as well as documented melting that has occurred from 1961 to 2003.

Estimated Maximum Sea-Level Rise from the Total Melting of Present-Day Glaciers		
LOCATION	VOLUME (CUBIC MILES/CUBIC KILOMETERS)	POTENTIAL SEA-LEVEL RISE (FEET/METERS)
East Antarctic Ice Sheet	6,247,136/26,039,200	213/64.80
West Antarctic Ice Sheet	782,595/3,262,000	26/8.06
Antarctic Peninsula	54,484/227,100	1.5/.46
Greenland	628,571/2,620,000	21/6.55
All other ice caps, ice fields, and valley glaciers	43,184/180,000	1.5/0.45
Total	7,755,972/32,328,300	263.5/80.32

Source: USGS

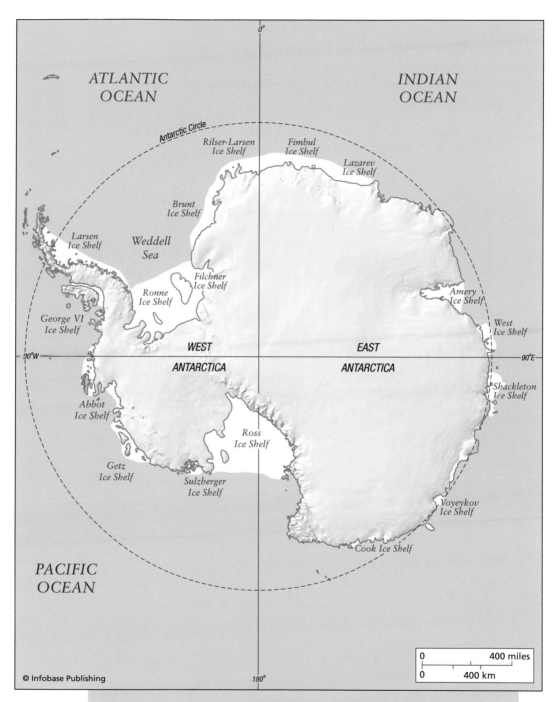

ATLANTIC
OCEAN

INDIAN
OCEAN

Antarctic Circle

Rilser-Larsen
Ice Shelf

Fimbul
Ice Shelf

Lazarev
Ice Shelf

Brunt
Ice Shelf

Larsen
Ice Shelf

Weddell
Sea

Filchner
Ice Shelf

Amery
Ice Shelf

Ronne
Ice Shelf

West
Ice Shelf

George VI
Ice Shelf

90°W

WEST

ANTARCTICA

EAST

ANTARCTICA

90°E

Shackleton
Ice Shelf

Abbot
Ice Shelf

Ross
Ice Shelf

Getz
Ice Shelf

Sulzberger
Ice Shelf

Voyeykov
Ice Shelf

Cook Ice Shelf

PACIFIC
OCEAN

© Infobase Publishing

180°

| 0 | | 400 miles |
| 0 | | 400 km |

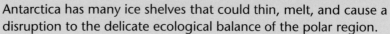

Antarctica has many ice shelves that could thin, melt, and cause a
disruption to the delicate ecological balance of the polar region.

Sea-Level Rise (inches/centimeters per year)		
SOURCE	1961–2003	1993–2003
Thermal expansion	0.02/0.05	0.02/0.05
Glaciers and ice caps	0.02/0.05	0.03/0.08
Greenland Ice Sheet	0.002/0.005	0.008/0.020
Antarctic Ice Sheet	0.006/0.015	0.008/0.020
Sum	0.04/0.10	0.1/0.25
Observed	0.07/0.18	0.1/0.25
Source: IPCC, 2007		

EFFECTS OF SEA-LEVEL RISE

The impacts of rising sea levels go beyond the world's coastlines. As global warming continues and sea levels rise, storm surges will increase in intensity, destroying land further inland from the coastal regions. Flooding will become one of the major problems, as well as several other negative impacts.

As ocean waters move inland, freshwater areas will become contaminated with salt water. As saline water intrudes on rivers, bays, estuaries, and coastal aquifers, they will become unusable. Wildlife that depends on freshwater will have its habitat negatively impacted, and drinking water will become unusable. Erosion will increase along coastlines. This will spell disaster for many of the world's population that currently live along the coasts. It will leave many people homeless and be economically devastating, especially in underdeveloped countries.

As wetlands, mangroves, and estuaries are impacted, fragile habitats will be lost around the world. Species will become threatened, endangered, and extinct. Other marine ecosystems will also be harmed, such as coral reefs. Reef habitats are extremely fragile, and significant physical changes in their environment can quickly destroy them.

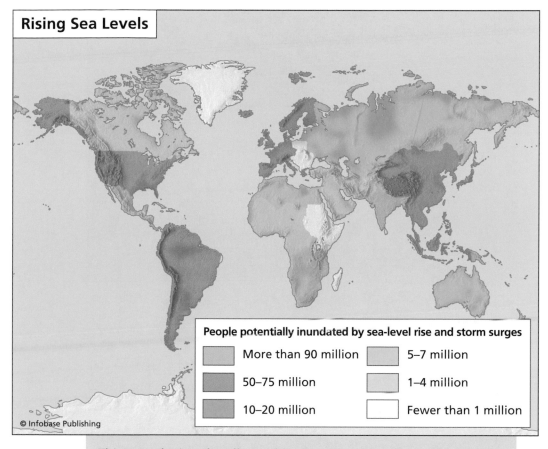

Rising Sea Levels

People potentially inundated by sea-level rise and storm surges

More than 90 million	5–7 million
50–75 million	1–4 million
10–20 million	Fewer than 1 million

© Infobase Publishing

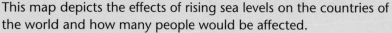

This map depicts the effects of rising sea levels on the countries of the world and how many people would be affected.

The most vulnerable areas are the low-lying countries of the world with extremely large coastal populations, such as Bangladesh, the Maldives, Vietnam, China, Indonesia, Senegal, Tuvalu, Mozambique, Egypt, the Marshall Islands, Pakistan, and Thailand. Developing countries do not have the economic resources to implement *adaptation* measures, such as building sea walls to hold back rising waters. If sea levels rise, the inhabitants of the coastal areas will have no other choice but to move inland to higher ground, if possible, losing what they have at lower levels. If mass migrations result, this could lead to a host of other negative issues, such as hunger, disease, and civil unrest.

Many locations worldwide that are low-lying islands will become especially vulnerable if sea levels keep rising. *(Nature's Images)*

Island states are particularly vulnerable. One of the nations most at risk is the Maldives. This nation lies in the Indian Ocean and is comprised of nearly 1,200 individual islands. Their elevation above sea level is only six feet (2 m). With a population of more than 200,000 people, if sea levels were to rise significantly, the entire country could become uninhabitable, leaving the population homeless. The Marshall Islands and Tuvalu, in the Pacific, face the same situation. Rising sea levels there would also first contaminate drinking water supplies, then drown the landmasses, leaving the population homeless. Other vulnerable locations include the cities of London, Amsterdam, Shanghai, and Jakarta, as well as many of the Caribbean islands.

In a study conducted by Sugata Hazra, an oceanographer at Jadavpur University in Calcutta, India, over the past 30 years, 31 square miles (80 km²) of the Sundarbans have disappeared because of rising sea level, displacing more than 600 families. Another area, Ghoramara, has had all but two square miles (5 km²) of its land submerged, which is now half the size it was back in 1969. The Sundarbans represent some of the world's biggest collection of river delta islands that lie between India and Bangladesh. Sea-level rise has contaminated the drinking water and destroyed the forested areas in the ecosystem. It has also threatened the existence of the wildlife, including the Bengal tiger. More than 4 million people live on the Sundarbans, a tiny island state, and hundreds of families have already been forced to move to refugee camps on neighboring islands. This is just one example of how rising sea levels are impacting developing countries today.

The impacts are not just limited to other countries; they will also be felt in the United States. Both the Atlantic and Gulf coasts face serious impact in the face of encroaching ocean levels and saline waters. Washington, D.C., is one of the more vulnerable areas. Higher sea levels would flood the Potomac and encroach on many famous historic landmarks. Baltimore and Annapolis are in a similar situation.

In the Mississippi delta, the loss of wetlands is a serious issue. Changes in sea level can cause wetlands to migrate landward. The Atlantic coast is one of the more sensitive areas to wetland vulnerability. Not only is this a problem for the natural habitat, but historically, these areas have been one of the most rich commercial fisheries in the world. If wetlands are endangered or destroyed, it would also have significant economic ramifications. These issues make the monitoring and control of rising sea levels a critical concern. Areas particularly in danger include Florida, Mississippi, Louisiana, North Carolina, South Carolina, Alabama, Georgia, and Texas.

FUTURE PROJECTIONS

One of the key issues is that of future sea-level rise. Because the ocean's thermal inertia is so great (it can hold large amounts of heat over long periods of time), it will take decades for the oceans to adjust their levels to the heat absorbed. In fact, for the heating caused by greenhouse-gas emissions already released into the atmosphere, sea levels are still trying to find a point of equilibrium. Therefore, even if all greenhouse emissions stopped today, there would still be a lag time for the oceans to stop rising. During this lag time, the oceans will likely rise another 5–12 inches (13–30 cm) by 2100. In the 2007 Fourth Assessment Report of the IPCC, a sea-level rise of 7–23 inches (18–58 cm) by 2100 was projected.

According to the USGS, based on information obtained from both tidal gauges and satellite measurements worldwide, scientists can say with confidence that sea-level rise has increased during the 20th century. Increased scientific knowledge has also clarified some issues that were not well understood previously, such as that the large polar ice sheets are far more sensitive to surface warming that initially thought, with significant changes currently being observed on the Greenland and West Antarctic Ice Sheets. Scientists now realize that these melting ice sheets can add water mass much more quickly to the oceans than previously assumed and play a significant part in overall global sea-level rise. Today also marks a notable consensus among specialists in climate change at USGS. It is largely recognized and accepted that there could be a rapid collapse of the polar ice sheets, and scientists have keyed in on the fact that anthropogenic actions, such as burning fossil fuels, could result in triggering an abrupt sea-level rise before the end of this century. They stress public education and political policy be brought to the forefront in order to deal most effectively with a situation that affects every person living on Earth now and in the future.

Abrupt Climate Change

One of the effects of climate change and global warming that concerns scientists today is the concept of abrupt climate change. Normally, significant changes that take place in the climate happen gradually over a long period of time—thousands of years. The climate does not always take this long to change, however, and when the changes occur in a matter of years to decades, this is an extremely fast rate. This chapter looks at what qualifies as abrupt climate change, evidence of past occurrences, and what triggers it. It then focuses on how the paleo evidence can be used to predict the future. Next, it examines the various warning signs associated with abrupt change and how likely it is that global warming could trigger a new episode. Finally, it clarifies various misconceptions that have been generated in the past concerning the topic.

ABRUPT CHANGES IN CLIMATE

In the past, it was believed that major climate change could only occur gradually, that the climate system was so huge and complex that any significant changes would take centuries or longer to occur. All that thinking began to change when scientists started drilling ice cores in Antarctica and Greenland. Hidden within the ice cores were detailed preserved histories of past climate going back 110,000 years in year-by-year layers. Located within these ice cores are bubbles of atmospheric gas, trapped in the ice, preserving a small piece of Earth's atmosphere at the time. Today, when scientists study the ice cores, they analyze these gas inclusions found in the cores and can accurately measure the chemical composition of Earth's atmosphere. For instance, they are able to see how much CO_2 existed in the atmosphere during different times of Earth's past. By measuring the amount of the "heavy" isotope of oxygen, as ^{18}O, they infer what the average atmospheric temperature was like. Water vapor with a greater concentration of ^{18}O molecules *condenses* out of clouds when the temperature is colder.

Through the study of ice cores, it has been determined that abrupt climate changes have occurred within decades in Earth's past. Just in the past 15,000 years, there have been various intensities of abrupt climate change events that have lasted for various periods of time.

According to the U.S. National Research Council, an abrupt climate change is when a climate system changes characteristics into a completely different mode so quickly that both humans and natural systems have a hard time adapting to it. This might include variations of temperature, precipitation, or extreme weather in a decade or in just years. Abrupt climate change also happens much quicker than it normally would under expected natural systems. Just as these episodes have occurred in the past, it is predicted that they will likely occur in the future.

If temperatures abruptly warm, some areas in the world may benefit, but many others may experience deadly heat waves. If areas drop in temperature, it could present problems for people in areas unaccustomed to cold temperatures, cause serious health issues, create economic hardships, and impact transportation corridors. If an area becomes too dry, it could destroy farms and the ability to grow food, limit drinking water availability, and spread disease. Any type of abrupt climate change could have a serious impact on civilization, as it has in the past with the destruction of great civilizations, such as the Mayan and Anasazi cultures.

According to the IPCC, human activity today is affecting climate, but it is difficult to tell the difference between human-induced changes and natural changes. Many scientists are trying to determine whether human influences can trigger abrupt climate change. Still difficult to predict, this is one of the areas that requires much research. Scientists do know, however, that abrupt changes will heavily impact both society and ecosystems. If major regional or global climate shifts were to happen today, there would be severe consequences for both humans and the natural environment.

It is such a serious issue that in 2002, the National Academy of Sciences—a board of scientists established by Congress in 1863 to advise the U.S. federal government on scientific issues—created a highly detailed report called "Abrupt Climate Change: Inevitable Surprises." It represents the most authoritative source of information about abrupt climate change available. One of the surprising findings from the analysis of ice cores was how many times Earth's climate had rapidly changed. In fact, an ice core taken from the Greenland Ice Sheet shows frequent sudden warmings and coolings of 15°F (8°C) or more, many of the changes occurring in less than 10 years. One of the most alarming incidents occurred 11,600 years ago at the end of the last ice age. This particular event, the Younger Dryas, showed that there was a 13°F (8°C) warming that occurred in less than a decade, which was also accompanied by a doubling of snow accumulation in three years. Most of the doubling occurred in just one year's time.

Scientists also found that these abrupt changes did not just affect Greenland but occurred worldwide. In fact, during the past 110,000 years, there have been at least 20 abrupt climate changes and only one period of stable climate—the most recent 11,000 years of modern climate. What this tells scientists is that today's climate is a brief pause in a long history of global climate change.

EVIDENCE OF PAST ABRUPT CLIMATE CHANGE

Scientists use various forms of proxy data as evidence of past abrupt climate change. The *proxies* most commonly used include tree rings, corals, speleothems (cave deposits), glacial deposits, polar ice caps, lake and marine sediments, and pollens. It was through the use of proxies (ice cores) that scientists were able to determine that the Younger Dryas had occurred around 12,800 years ago. The event was an unusual break in the climate cycle of Earth's warming up after the Ice Age. The Earth returned to a glacial-like climate and persisted as a glacial ecosystem for the next 1,200 years. Then, around 11,600 years ago, the climate phase abruptly ended, and the temperature increased 17°F (10°C) in just a 10-year interval.

There have been other events over the past 50,000 years. The Heinrich events occurred during the last ice age and involved multitudes of icebergs broken from glaciers that melted and added freshwater to the ocean. This freshwater was believed to have played a role in moderating the circulation of the Atlantic Ocean and causing a sudden cooling in the climate. The Dansgaard/Oeschger events were rapid fluctuations of temperature that also occurred during the last ice age. Scientists have recorded 23 fluctuations that occurred 23,000 to 110,000 years ago where the atmosphere warmed up quickly, then cooled down over longer periods of time.

Another episode, called the 8,200-year event, lasted only 100 years, but temperatures in the North Atlantic region dropped an average of 5°F (3°C) and caused widespread dry conditions. Scientists speculate this was caused by freshwater being added to the Atlantic and slowing the thermohaline circulation. More recently, the interval known as the medieval warm period was a time of abrupt warming (0.3°F, or 0.2°C) about 1,000 years ago. This period ended 700 years ago with the start of the Little Ice Age, another abrupt climate change. The Little Ice Age existed from the 1300s to the mid-1800s. Its cooling was not as severe— only about 1.7°F (1°C).

Abrupt climate change has also been documented by archaeologists and climatologists as one of the chief causes for the fall of great civilizations. Among these are the Maya, the Anasazi, the Akkadian Empire, and civilizations in Mesopotamia, Egypt, India, and China, as well as cultures in Europe such as the Vikings.

CAUSES OF ABRUPT CLIMATE CHANGE

One of the overwhelming causes of abrupt climate change that scientists focus upon is the sudden shutdown or start-up of the meridional overturning circulation (MOC). Also commonly referred to as the thermo-haline circulation, or *great ocean conveyor belt,* it is a complex network of ocean currents in the Atlantic. The Gulf Stream, which is the current that transports a significant amount of heat northward from Earth's equatorial region toward western Europe, helping to warm its climate, is part of the circulation. In fact, if it were not for the Gulf Stream, the North Atlantic and Europe would be 8°F (5°C) cooler, especially in the winter. If this extensive current were to shut down, it would negatively impact the entire ocean-atmospheric system and cause adverse effects worldwide in not only ocean circulation but also the *jet stream* in the atmosphere that drives storm systems.

Based on evidence retrieved from ice cores in Greenland, scientists have determined that the MOC has been shut down in the past and that every time it has been shut down, an abrupt climate change has occurred. The chief mechanism for shutting down the great ocean conveyor belt is the addition of freshwater. In the circulation process, the high-density water sinks and the low-density water rises. Salty water is denser than warm water; therefore, combining these two principles, a place where cold, salty water exists is where the water sinks vertically in the ocean.

In the tropical regions of the Atlantic Ocean where there is a tremendous amount of heat, large amounts of water get evaporated. This area contains warm, salty water that flows westward toward North America, travels along the east coast of the United States, and then heads northeast toward Europe as the Gulf Stream current. Once it reaches Greenland, it cools and sinks to the bottom of the ocean. At this point, the current heads south again and travels toward Antarctica and around to the Pacific Ocean. A water molecule travels through the entire circuit of the great ocean conveyor belt in about 1,000 years.

Abrupt climate change could occur if freshwater were added near Greenland where the Gulf Stream cools and sinks. Freshwater can be added through the melting of glaciers, a situation that is becoming more probable with global warming. If too much freshwater is added, the water is no longer dense enough to sink, which then causes the circulation to slow down or stop all together. If the circulation stops, an abrupt climate change will be triggered. One of the unknowns in this case is how much freshwater would have to be added in order to turn off the conveyor belt. The circulation was shut off during the Younger Dryas event and then suddenly restarted after 1,100 years. Even though scientists have developed computer models to study the phenomenon, they have not been able to duplicate the start-up or shutdown processes, meaning they still have much to learn about the natural process itself.

Today, as Earth heats up with global warming, there could be an increase in precipitation in some areas, as well as a melting of freshwater ice in the Arctic Ocean. Scientists are concerned this would dilute the Gulf Stream. According to the Environmental Literacy Council, over the past 40 years, salinity measurements within the North Atlantic have shown that the area has been decreasing in its level of salinity. They caution that if it continues and the Gulf Stream were to eventually come to a halt, western Europe and eastern North America would cool up to 8°F (5°C)—about the same difference as the global average temperature difference between today and the Ice Age.

Another cause of abrupt climate change centers on the cryosphere. This involves the interaction of the incoming solar radiation and the ice cover on Earth's surface. When large areas on Earth's surface are covered with ice, much of the incoming sunlight is reflected back into space, a property called "albedo." Scientists think this may cause a positive feedback situation for an abrupt cooling. As solar radiation is reflected back into space, it lowers the temperature on Earth, which encourages more snow and ice to form, which increases the albedo, continuing to accelerate the cooling of the climate.

A third mechanism of triggering climate change involves methane hydrates. Methane is a greenhouse gas, but in comparison to carbon dioxide, it is a much more powerful absorber of heat. Because of this, even small releases of methane in the atmosphere could have enormous effects on Earth's climate. Methane hydrates are solid compounds of methane and water and are mainly located in the Arctic *tundra* and in marine sediment. It is believed that release of methane hydrates in the past have caused abrupt climate changes. Today, there is concern that if there is an increase in temperature and the Arctic ice continues to melt, the methane hydrates (currently buried under the Arctic tundra) would be released into the atmosphere, heat it quickly, and cause an abrupt climate change.

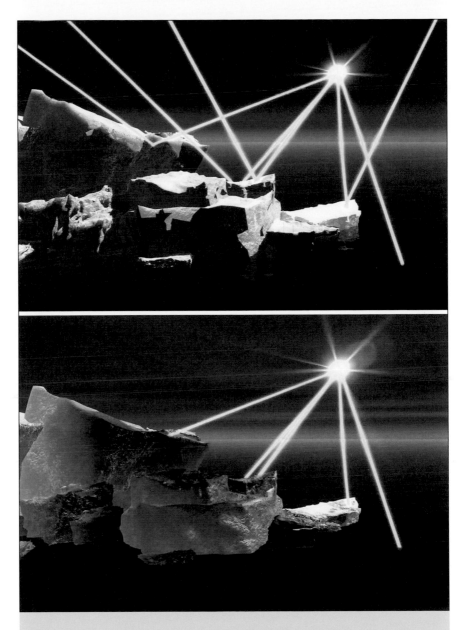

Top: When snow and ice cover the planet, the albedo is high. As incoming solar radiation reaches the Earth's surface, it is reflected off the ice and snow and sent back into the Earth's atmosphere, keeping the Earth's surface cool. This can serve as a positive feedback, encouraging more snow and ice to form in the cool environment. *Bottom:* If the ice on the surface of the Earth begins to melt or becomes covered with soot and other pollutants, it exposes darker surfaces. The incoming solar radiation heats up the darker surface, causing more snow and ice to melt. This starts a cycle of increased melting, referred to as negative feedback. *(NASA)*

USING PALEO EVIDENCE TO PREDICT THE FUTURE

Understanding Earth's past climate behavior provides priceless insight into what it may do in the future. Long winters and glacial advances are recorded in proxy data from glaciers, sea ice, and soil samples. Deep-sea sediment cores show that icebergs traveled as far south as the waters off the coast of Portugal.

Scientists at NOAA use paleoclimatic data to identify abrupt climate changes. These intervals have occurred as global events and as local events; some severe, others less severe. NOAA experts have studied events that occurred hundreds to tens of thousands of years ago. During the Holocene (the past 11,000 years), several climate changes have been centered on drought events; many of these well-documented events coincided with the fall of great civilizations.

One thing scientists at NOAA have concluded is that from the existing proxy data uncovered so far, the climate has changed in the past much more rapidly and intensely than anything humans have seen during their existence on Earth. Many of the most abrupt climate changes have involved major changes in the ocean-atmospheric-ice-land systems.

Scientists test their theories against existing data, partly through the creation of models. Thus far, they have determined that Earth's climate system has two different stable modes, which it switches rapidly between. Over the last few hundred thousand years, the paleo record shows warm interglacial stages, which can last for 10,000 years or more, interspersed with glacial stages. Each of the glacial cycles on Earth is similar in terms of temperature and duration. Scientists would like a better understanding, however, of which mechanisms trigger the abrupt changes.

One thing they have a good understanding of is rapid ice-sheet melting in the transition from cold glacial periods to warm interglacial periods. This is supported by observation. Experts at NOAA are currently working on computer models and improving the algorithms to simulate climate change. Scientists there are also debating other causes of abrupt change. One issue that is highly debated concerns the atmospheric carbon dioxide levels that accompany the glacial-interglacial cycles. These are still not well understood and are viewed as an important piece of knowledge to have in order to restrict abrupt changes before they happen in the future and in order to plan for and protect society from the resulting negative impacts.

WARNING SIGNS TODAY

Scientists caution today that it is possible for abrupt climate change to occur simultaneously with global warming. In fact, according to the Woods Hole Oceanographic Institution, global warming is a destabilizing factor that makes abrupt climate change even more probable. A report issued by the U.S. National Academy of Sciences (NAS) stressed that "available evidence suggests that abrupt climate changes are not only possible but likely in the future, potentially with large impacts on ecosystems and societies."

Scientists have recognized the need to understand more about the dynamics of the ocean-atmosphere system. In particular, the great ocean conveyor belt needs to be studied and better understood so that when certain thresholds are reached, there will be ample warning concerning climate change.

One warning sign scientists have about the conveyor belt is new data showing that since the mid-1960s, the subpolar seas that feed the North Atlantic have progressively become less salty to depths of 3,280–13,123 feet (1,000–4,000 m). This represents the largest oceanic change ever measured in the timeframe of modern instruments.

Experts at Woods Hole have confirmed that signs of a possible slowdown already exist. In particular, the flow of cold, dense water from the Norwegian and Greenland areas in the North Atlantic has decreased by 20 percent since 1950. They are trying to understand better the ocean conveyor belt, especially the deep ocean currents.

Climatologists have just started to measure deep ocean water properties and currents at critical locations with long-term moored buoy arrays. The ocean is so vast, however, that huge expanses are still unmonitored. The future goal of the U.S. Global Change Research Program is to develop new techniques to assist in measuring cycles in the ocean in order to predict and identify abrupt climate change. Some warn that learning as much as possible about the mechanisms of abrupt climate change is important for future progress. If the mechanisms of climate change are better understood, then society's abilities to monitor, plan for, and react to changes will be more efficient. Contingency plans should be in place to handle impacts to water resources, agriculture, energy, transportation, and public health.

GLOBAL WARMING AS A TRIGGER

According to Dr. Richard B. Alley, a professor of geosciences at Pennsylvania State University and an associate at Earth System Science Center at Pennsylvania State University, in an article in *Scientific American,* "Humans are pushing certain aspects of climate closer to the thresholds that could unleash sudden changes [in climate]." If global warming triggers climate change, humans will be presented with many challenges. Although it is true that some regions will become more habitable with warmer temperatures, such as at higher latitudes, where it is expected that farming may be possible (if the right type of soil exists for agriculture), other regions will become so warm that heat waves will wreak havoc, threatening agriculture, the availability of water resources, and the health of humans and ecosystems. Likewise, climate changes to cold climates could threaten the health of humans and ecosystems and disrupt transportation and economic systems.

The discoveries that scientists made while studying ice cores from Greenland in the early 1990s are what gave scientists an appreciation for the concept of abrupt climate change. It also made them realize that because of Earth's ability to shift climates within a decade or less, global warming needs to be taken very seriously. It could be a critical factor pushing Earth's climate even faster toward a sudden shift than what would normally be expected.

Scientists studied cores of ice that are up to 1.9 miles (3 km) long to obtain a climate history spanning 110,000 years. What they discovered was startling: There was evidence locked in the ice core revealing a detailed history of many wild fluctuations in climate. There were many long cold periods when the temperature dropped 10°F (6°C) in just a few years, interspersed with brief warm periods. They also found that since the Ice Age 11,500 years ago, the climate gained half of its heat—more than 17°F (10°C)—in just 10 years. This episode correlates with an increase in precipitation in other areas of the world. Because

There will be impacts to ecosystems in the event of abrupt climate changes. If the ice caps were to melt in the Antarctic, it would have a negative impact on penguin habitat. *(NOAA)*

of this, large amounts of methane were released into the atmosphere as wetlands flooded the Tropics and thawed in the warming northern latitudes.

They found that there were more than 20 intense, abrupt warming episodes, followed by cold periods, only to be repeated again by another warming that only took a few years to occur. Through proxy data, they were also able to determine that the cold, wet episodes in Greenland correlated with cold, dry, windy conditions in North America and Europe. Conditions in the South Atlantic and Antarctica, however, were extremely warm. Also through the analysis of proxy data, they discovered that 5,000 years ago a sudden dry spell converted the Sahara from a lush, green landscape to the dry, hostile environment it is today.

Climate scientists give us a warning as a result of their research on abrupt climate change. According to Alley, crossing a climate threshold is similar to tipping a canoe. In a canoe, a passenger can tip it and be safe initially, but as the leaning progresses to a certain point, it hits a critical threshold that flips the canoe over to the point of no return. Just as with the canoe, the climate system is robust but also has its limits. Once pushed too far, there comes a point where a climatic shift is inevitable and unstoppable. The key is to recognize what pushes the climate to that point and avoid creating situations that trigger them. The following table illustrates what can cause a climate to cross a particular threshold and trigger a climate change.

Alley and most climate experts feel that humans today are tempting abrupt climate change by interfering with so many aspects of the natural world, in particular, the human-induced increases in atmospheric concentrations of greenhouse gases, put there by the excessive burning of fossil fuels. Because of the human influence, the IPCC has predicted that global temperatures will rise 2.5–7.5°F (1.5–4.5°C) in the next 100 years. This warming is what could add freshwater to the North Atlantic, causing the North Atlantic conveyor to slow down or halt, cooling Europe and the eastern United States.

Abrupt climate change is considered an issue of national security. A report was prepared for the Pentagon on the implications of abrupt climate change toward national security and the possibility of human-caused global warming leading to the collapse of U.S. national security specifically. The Pentagon predicts a near-term collapse of the "Atlantic Ocean conveyor belt." If it were to completely shut down it could change precipitation amounts and negatively impact agriculture and populations. The rapid cooling in Europe, diminished rainfall amounts in many key agricultural and urban areas, along with the negative impacts to urban areas and the consequent disruptions in food and water supply pose enormous security issues. This, in turn, would cause significant geopolitical and security issues for countries to deal with, including the possibility of war.

This situation is serious enough that the U.S. Department of Defense created a think tank called the Global Business Network to assess the national security implications of a total shutdown of the North Atlantic conveyor. The think tank considers the situation very serious. In its report, the Global Business Network states: "Tensions could mount

Crossing Climatic Thresholds			
CLIMATE DRIVER	THRESHOLD CROSSING	RESULTING CLIMATE SHIFT	SOCIAL CONSEQUENCES
The North Atlantic conveyor current carries warmth northward from the Tropics to Europe.	The addition of freshwater in the North Atlantic slows or stops the conveyor current.	Temperatures fall in Europe and the eastern United States.	It becomes difficult to produce agriculture, and transportation is interrupted.
Evapotranspiration from plants provides the moisture for much of the rain received in farming areas.	A dry spell stresses and kills many plants, decreasing available moisture in the air and causing the area to become continually drier and unproductive.	A mild dry spell can be turned into a severe, long-term drought.	Land cannot be farmed, people cannot grow food and go hungry, health issues become critical, and the economy is negatively impacted.
Currents in the Pacific Ocean control temperature and weather.	Natural phenomena, such as El Niño, change sea-surface temperatures.	Weather patterns on land shift, causing either unplanned violent storms or drought.	Croplands can dry up, while other areas can be flooded.

(Source: Richard B. Alley. "Abrupt Climate Change." Scientific American November 2004, pp. 62–69.)

around the world. Nations with the resources to do so may build virtual fortresses around their countries, preserving resources for themselves. Less fortunate nations may initiate in struggles for access to food, clean water, or energy."

One thing to keep in mind, according to Alley, is that rapid changes in climate are not easily predictable. No one currently knows what the exact critical threshold is that can flip the climate to change abruptly. Instead, it can be better compared to conducting a large-scale science experiment in which humans are experimenting on Earth in a real-time laboratory experiment. The best scientists can do right now is study the past, observe the present, and develop models that fit both time intervals so that they can project them toward future scenarios.

THE DAY AFTER TOMORROW

The Hollywood science-fiction film *The Day After Tomorrow* (2004), an action-packed thriller about abrupt climate change, caught the attention of moviegoers around the world. But was this film science or science fiction? According to experts at the Ocean and Climate Change Institute and Woods Hole Oceanographic Institution, the abrupt climate change depicted in the movie over a time span of just a few weeks could not happen that fast. The climate is a huge, complex interactive system involving the land, atmosphere, and oceans, and a change in ocean circulation that could trigger a large-scale climatic change would take decades. Not only can the oceans not change that fast, but also ice sheets and glaciers cannot melt in a few weeks, either.

Most of the references to abrupt climate change to scientists revolve around the disturbance of the Atlantic conveyor belt and its key impact on the North Atlantic, Europe, and the eastern United States. A cooling in these areas, however, would not cause a global cooling or a global ice age.

Scientists predict that the degree of cooling that would occur would depend on the changes in ocean circulation. If changes were to occur between now and the next 50 years, the cooling effect may be more severe than if ocean circulation changes did not happen for 50 to 100 years. The reason for this is that the longer that global warming continues to rise, its warming effects may counterbalance and then outweigh the cooling effects.

Climate experts also commented on the hurricane that swept down from the Arctic in the movie. Hurricanes cannot form in the Arctic because hurricanes need heat and moisture from warm ocean waters to grow. The immense size of the storm is also impossible. No single storm system could reach the size of the Northern Hemisphere; the pressure gradients needed work on much smaller scales.

Nor is there any need to worry about the -150° wind chill, either, they say. The lowest temperature ever recorded on Earth was in Vostock, Antarctica, on July 21, 1983: a bone-chilling -128.6°F (-89.2°C).

Alley also stresses that there may be certain climate thresholds in existence that experts have not yet discovered. Scientists do agree that unexpected changes could be devastating and have cautioned for global

Furthermore, storm surges do not arrive in the 40-foot (12-m) variety in New York, nor do they arrive as a single, giant wave. The wave in the movie more closely resembled a tsunami. Storm surges usually occur as a rise in water over a time period of several hours. According to experts, models produced by NOAA have predicted that storm surges of 20 feet (6 m) are possible in New York if a powerful hurricane hit northern New Jersey.

Mammoth chunks of ice breaking off of Antarctica do not cause tidal waves. In fact, when an ice shelf detaches from Antarctica, it does not even cause a rise in sea level because it is already floating in the ocean. In 2002, a piece of the Larsen B Ice Shelf the size of the state of Rhode Island broke off and did not cause a tsunami or any rise in sea level.

Global warming, however, does contribute to a rise in sea level in two different ways: 1) when ice on land melts and flows into the ocean, and 2) when water is heated, because it expands. This is called "thermal expansion" and is more pronounced in the mid- to polar latitudes. If the West Antarctic Ice Sheet—a huge mass of ice roughly the size of Alaska and California combined—were to melt into the ocean (a process that could take thousands of years at today's rate of global warming), sea levels could rise about 20 feet (6 m).

The bottom line is that experts do acknowledge abrupt climate change as a real phenomena that can take place in a matter of decades or less. The present state of global warming is melting the glaciers and ice sheets around the world, which is contributing to the freshening of the ocean water, most notably the North Atlantic. As far as shutting down the North Atlantic conveyor, many climatologists are in the process of researching and modeling it, in search of a reliable prediction of what Earth's climate will be like—the day after tomorrow.

Source: Woods Hole Oceanographic Institution

reductions of carbon emissions as a way to slow global warming. Further research is also needed, as it takes time to understand the extreme complexity of the climate system.

The U.S. National Research Council has recommended that societies also need to develop plans to be able to cope with abrupt climate change before the next climate "surprise" occurs. They have suggested that communities plant trees in order to stabilize soil in the event of drought so that winds do not carry away dry soil, as they did in the dust bowl of the 1930s in the U.S. Midwest. Regions should also plan for and manage their water resources wisely so that if a drought occurred, communities would be prepared.

MISCONCEPTIONS ABOUT ABRUPT CLIMATE CHANGE

Even though scientists have uncovered a great deal of information about the climate system and how it works, there are still some uncertainties, and because of them, some misconceptions. Some have stated that abrupt changes in climate are natural, that they have occurred throughout history. The reality is that while natural climate change has occurred and still does, a large component today is attributed to human interference. According to the American Geophysical Union, "It is scientifically inconceivable that—after changing forests into cities, putting dust and soot into the atmosphere—humans have not altered the natural course of the climate system." Greenhouse gases are being added at an alarming rate due to the burning of fossil fuels and deforestation.

Another misconception is that the ocean does not play an important role in climate. The ocean stores roughly 1,000 times more heat than the atmosphere; but because the atmosphere has the capability to move the heat more quickly, they equal each other out. The ocean does play a critical role in transferring heat; it is what keeps Europe warm in the winter, for example.

There is also some confusion when scientists talk about global warming on one hand and a cooling of Europe on the other hand. Although they seem contradictory, they are not. As global warming melts ice cover in the Arctic, it adds freshwater to the North Atlantic, which could threaten to slow, or stop, the North Atlantic conveyor belt, which supplies warmth to Europe, thereby putting the continent into a cold freeze. But it would not completely stop the Gulf Stream. It is also important to note that although global warming could lead to regional cooling such as this, it cannot cause a global ice age. Physical conditions are different now compared to the last Ice Age. For instance, the shape of Earth's orbit is different now than it was during the last Ice Age, as is the tilt of Earth's axis. There is also a much higher concentration of CO_2 in the atmosphere.

Global warming will affect the *hydrologic cycle,* however. As global warming heats the atmosphere, more water is evaporated, which traps additional heat. In theory, the hydrologic cycle should accelerate as a result of global warming, which will then accelerate the process. There is still a lot of uncertainty about the role of the hydrologic cycle, however. Many models have been developed, and a wide variety of results have been obtained.

According to Environment News Service, a study conducted by Ruth Curry of the Woods Hole Oceanographic Institution; Bob Dickson of the Centre for Environment, Fisheries, and Aquaculture Science in Lowestoft, United Kingdom; and Igor Yashayaev of the Bedford Institute of Oceanography in Dartmouth, Nova Scotia, Canada, found that an acceleration of Earth's global water cycle—caused in part by global warming—may affect global precipitation patterns that determine the distribution, severity, and frequency of droughts, floods, and storms. They caution that a water cycle acceleration would intensify global warming by adding more water vapor to the atmosphere, which could continue to freshen the North Atlantic Ocean waters to the point that it could disrupt the ocean conveyor belt, triggering abrupt climate change.

Elise Ralph, associate director of the NSF (which funded the research) said, "This study is important because it provides direct evidence that the global water cycle is intensifying. This is consistent with global warming hypotheses that suggest ocean evaporation will increase as Earth's temperature does." Their models showed that the properties of Atlantic water masses have been changing radically over the past 50 years. The water at the poles is becoming fresher, while the water, in the tropical oceans is becoming saltier. The scientists involved in the study said that "these results indicate that freshwater has been lost from the low latitudes and added at high latitudes at a pace exceeding the ocean circulation's ability to compensate."

The scientists also noted that an accelerated water cycle is causing increasing rain and snow in higher latitudes, which is also contributing to the freshening of the North Atlantic Ocean waters. As a result of their study, the scientists stated: "Monitoring Earth's hydrological cycle is critical because of its potential near term impacts on Earth's climate."

In general, scientists know that in polar areas, the surface and deep waters have been increasing in freshwater for the past 40 years. In the Tropics, surface waters have been losing freshwater due to increased evaporation. These conditions could cause an accelerated hydrological cycle.

Tropical Cyclones and Other Severe Weather

This chapter examines the effects of tropical cyclones and other extreme weather. It first looks at hurricanes and their connection with global warming and focuses on human activity that may be contributing to hurricanes. It presents storm surges, what they are, and why their prediction is important, as well as what scientists have to say about the trend of hurricanes in 2004 and 2005. Finally, this chapter touches on current research being conducted on hurricanes, tornadoes, and thunderstorms.

HURRICANES AND GLOBAL WARMING

A hurricane is an intense *tropical storm* in which sustained wind speed exceeds 74 miles (119 km) per hour. According to the 2007 Fourth Assessment Report of the IPCC, it is "more likely than not" (meaning better-than-even odds) that there is a human contribution to the observed trend in the rise of hurricane intensity since the 1970s. The IPCC states that "it is likely (more than 67 percent odds) that future tropical cyclones (hurricanes) will become more intense, with larger peak wind speeds and heavier rain associated with continually rising sea surface temperatures."

Scientists at NOAA recognize two factors that contribute to more intense hurricanes: ocean heat and water vapor. They also recognize that these two factors have increased over the past 20 years because of human activities, such as burning fossil fuels and deforestation. Both activities have significantly raised the CO_2 levels in the atmosphere.

They also warn that the world's oceans have already absorbed 20 times as much heat as the atmosphere over the last 50 years, which has warmed the waters to depths of 1,500 feet (457 m) in places. As the oceans warm, the water expands. Atmospheric humidity over the oceans has risen 4 percent since 1970, and because warm air holds more water vapor than cold air, this explains an increase in air temperature. They believe this is a visual, measurable result of global warming.

Another impact from this is higher storm surges, the height and amount of water that washes up on shore during a storm. Rising sea-levels mean that storm surges will be higher and more destructive, causing coastal flooding and erosion. Many people who live along coastal areas will have their property damaged and lost. The following table illustrates the resultant storm surge of different category hurricanes.

Saffir-Simpson Hurricane Scale		
CATEGORY	WIND SPEED (MILES PER HOUR/ KILOMETERS PER HOUR)	STORM SURGE (FEET/METERS)
1	74–95/119–153	4–5/1.2–1.7
2	96–110/154–177	6–8/1.8–2.5
3	111–130/178–209	9–12/2.6–3.8
4	131–155/210–249	13–18/3.9–5.5
5	>155/250	>18/5.5

Beach erosion is a serious problem. This erosion was caused by a northeaster on the Outer Banks, North Carolina. *(Richard B. Mieremet, Senior Adviser, NOAA)*

Some scientists believe that the strength of a storm and the length of time it lasts may be increasing as global warming emissions increase in the atmosphere. Rising sea levels intensify the damage along coasts from storms. Warmer ocean temperatures (>80°F, or >27°C) signify the potential for more hurricanes. Because natural cycles cannot explain recent warming trends in the oceans, some scientists are focusing more on global warming because the CO_2 levels in the atmosphere are much higher than they have been in the past 400,000 years.

As reported on MSNBC on June 16, 2005, Kevin Trenberth of the National Center for Atmospheric Research claims that warmer oceans and increased moisture could intensify the showers and thunderstorms that fuel hurricanes. He said: "Trends in human-influenced environmental changes are now evident in hurricane regions. These changes are expected to affect hurricane intensity and rainfall, but the effect on hurricane numbers remains unclear. The key scientific question is how hurricanes are changing."

Chris Landsea of NOAA counters by saying there is evidence for natural swings between high and low hurricane activity that extend for 25 to 40 years. He comments: "The last 10 years have been busy for the U.S.—similar to what we experienced between the 1920s and 1960s."

As reported by National Geographic News on August 4, 2005, a study run by Kerry Emanuel, a professor of atmospheric science at the Massachusetts Institute of Technology, that appeared in the journal *Nature,* found that hurricanes and typhoons have become stronger and longer lasting over the past 30 years, and that these upward swings correlate with a rise in sea surface temperatures. Kerry also concluded that the duration and strength of hurricanes have increased by about 50 percent over the last three decades.

According to the Union of Concerned Scientists, a study conducted in 2005 determined that over the past 30 years, hurricanes' destructive power has increased approximately 70 percent in both the Atlantic and Pacific Oceans. Another study in 2005 by P. J. Webster, J. A. Curry, and H. R. Chang of the Georgia Institute of Technology and G. J. Holland from the National Center for Atmospheric Research in Boulder, Colorado, determined that the number of hurricanes classified as category 4 or 5 (based on satellite data) has increased over the same period. These findings correlate with the rise in observed sea surface temperatures in regions where tropical cyclones usually originate.

In 2004, Ruth Gorski Curry of the Woods Hole Oceanographic Institution presented a study on hurricanes and climate change and observed that in both the Eastern and Western Hemispheres, tropical storm activity had been well above normal for several years. In addition, in March 2004, the South Atlantic experienced its first hurricane ever recorded. She cited several factors that have helped fuel the increased number and intensity of tropical storms in recent years:

- the absence of El Niño conditions in the Pacific
- stratospheric winds in the equatorial belt
- a wet season in the African Sahel
- unusually warm water temperatures in the Atlantic, western Pacific, and Indian Oceans

Tropical cyclones, such as this one off the coast of Brazil, are some of the most deadly storms on Earth. *(NOAA)*

Both climate models and observations support the idea that as global temperatures rise, the oceans are getting warmer. The models also agree that greenhouse warming will enhance the frequency and intensity of hurricanes in the coming century. T. R. Knutson and R. E. Tuleya at NOAA have looked at the potential for future storm trends and determined that if there is a 1 percent annual increase of CO_2 concentrations over the next 80 years, it would produce more intense storms and the amount of rainfall would increase 18 percent.

One major problem with rising ocean temperatures is that it will fuel hurricanes. Under ordinary conditions, the surface winds will churn up cooler water from the ocean, which will help slow the storm, but if the subsurface is too warm, it will only fuel the storm.

According to *USA Today*, experts had this to say about global warming:

- Kerry Emanuel, Massachusetts Institute of Technology: "Storms are lasting longer at high intensity than they were 30 years ago. Hurricane reported durations have increased by about 60 percent since 1949, and average peak storm wind speeds have increased about 50 percent since the 1970s.
- The scientists who support the link between global warming and hurricanes believe that a warming world has caused the oceans to heat up over the past several years, which is causing an increase in the number and intensity of hurricanes.
- The number of category 4 and 5 hurricanes has increased sharply over the past few decades, from about 11 per year in the 1970s to 18 per year since 1990. Rising sea surface temperatures encourage stronger hurricanes because they draw energy from warm ocean waters and release it in huge storms.
- Scientists believe that about half of the ocean's extra warmth in 2005 was due to global warming.
- According to the IPCC in one of its reports, the scientists said that during the 21st century, hurricanes, typhoons, and Indian Ocean and South Pacific cyclones are likely to produce higher winds and heavier rain in some areas, but there is no way to tell whether the frequency and locations of these storms could change.

A 2005 study by Kerry Emanuel published in *Nature* suggests that storm intensity and duration is linked to global warming and recent ocean warming trends. Using a measurement called the power dissipation index (PDI), which measures how destructive a storm is, Emanuel and his team came to the conclusion that in the last 30 years, the destructive power of storms has doubled in the Atlantic and Pacific Oceans, the bulk of this destruction occurring in the last 10 years.

STORM SURGE PREDICTION AND SIMULATION

Because so many of the coastal areas of the world are developed and populated with dense urban areas, there is an extreme risk to loss of

HOW HURRICANES ARE NAMED

Hurricanes are given names from lists that have been previously prepared. The names on the list are approved by the World Meteorological Organization (WMO) or by national weather offices. If a hurricane is extremely destructive, its name will be retired and never used again for a hurricane. For example, *Katrina* will not be used again.

In the North Atlantic and northeastern Pacific, both male and female names are alternated in alphabetical order during a season. The gender of the season's first storm alternates from year to year. Six individual lists are prepared ahead of time, and one list is used for each of the next six years.

In the North Atlantic, five letters are not used: *Q, U, X, Y,* and *Z*. In the northeastern Pacific, two letters are not used: *Q* and *U*. This means that there can be 21 possible hurricanes in the North Atlantic and 24 names in the northeastern Pacific before the names on the list run out. The only time a region used up its entire list was the North Atlantic in 2005. When the names are used up, additional names can be added to the list using letters from the Greek alphabet.

life and property damage if a hurricane comes ashore and storm surges flood coastal areas. Because of the extreme danger to life and the fact that it costs so much to clean up after the devastation of a hurricane, it is vital to take measures to avoid contributing to increases in temperature and global warming. Hurricane Andrew, for example, had 64 fatalities and cost $43.7 billion; Ivan had 124 fatalities and cost $14.2 billion; and Katrina had 1,836 fatalities and cost $125 billion.

The Federal Emergency Management Agency (FEMA), part of the U.S. Department of Homeland Security (DHS), is tasked with the primary mission to reduce the loss of life and property and protect the nation from all hazards, including natural disasters. During disasters such as hurricanes, FEMA works with other organizations that are part of the nation's emergency management system, such as local emergency management agencies and the American Red Cross. They are involved in work dealing with storm surges and flooding.

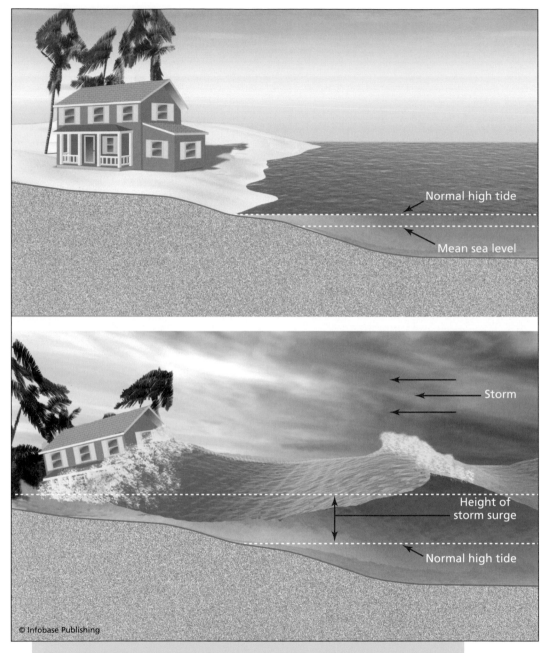

Storm surges happen over a period of time, as the ocean level continues to rise, often to the point of doing severe damage to structures in coastal areas.

New Orleans, Louisiana. Storm surge, Hurricane Katrina *(Jocelyn Augustino, FEMA)*

Computer models have been developed for coastal communities that can project various water saturation heights onto a coastal area for different hurricane categories. Computers can also simulate slow-moving and rapidly moving hurricanes in order to see the real-time potential effects of the storm. These types of computer simulations allow emergency planners to prepare evacuation and emergency preparedness plans.

GLOBAL WARMING AND THE HURRICANES OF 2004–2005

In accordance with a study in 2005 involving the Massachusetts Institute of Technology, scientists found that the destructive potential of tropical storms has doubled over the past 30 years. The hurricanes of 2004 and 2005 were right in line with what scientists warning about global warming have predicted. Some scientists have been predicting for a long time that extreme weather events such as hurricanes will increase as a consequence of global warming.

In a review printed by the *Times* on September 16, 2005, Peter Webster of the Georgia Institute of Technology analyzed all the records of hurricanes and typhoons since 1970 and said this: "What we found was rather astonishing. In the 1970s, there was an average of about 10 Category 4 and 5 hurricanes per year. Since 1990, the number of Category 4 and 5 hurricanes has almost doubled. Our work is consistent with the concept that there is a relationship between increasing sea surface temperature and hurricane intensity." This belief has also been voiced by the IPCC in its 2007 report that research confirms the North Atlantic's recent fiercer hurricanes are correlated with increases of tropical sea surface temperatures.

Scientists at the National Hurricane Center predicted that the 2004 hurricane season should experience above-normal hurricane activity. They predicted 12 to 15 named storms, of which seven would most likely reach hurricane strength. The end result was above what they predicted: 15 named storms, nine of which were hurricanes. This was the year of Hurricanes Charley, Frances, Ivan, and Jeanne.

The National Hurricane Center expected even more in 2005. It predicted 18–21 tropical storms and 9–11 hurricanes, five to seven of which were expected to reach a category 3 or greater. (Categories range from 1 to 5, 5 being the most destructive.) The results were significantly greater. The season of 2005 turned out to be the most active year on record. According to the National Hurricane Center, this is what happened.

- There were 28 tropical storms (the previous record was 21 storms in 1933).
- Of these, 15 developed into hurricanes.
- It was the first time that four named tropical storms formed before July 5 (Arlene, Bret, Cindy, Dennis).
- It was the first time that two category 4 hurricanes had formed before July 14 (Dennis on July 4–7 and Emily on July 10–16).
- It was the year of the most powerful hurricane ever recorded in the Atlantic Basin (Wilma).

- It was the year of three of the six most powerful hurricanes of all time in the Atlantic.
- It marked the first time that three category 5 hurricanes ever occurred in the same year in the Atlantic.
- It was the year of the most destructive hurricane in U.S. history (Katrina).
- It tied the record for the latest date of tropical storm formation. Zeta formed on December 30 (the previous record was set in 1954).

The 2005 Atlantic hurricane season was the most active season ever recorded.

CYCLIC THEORY

Some climatologists view the incidences of hurricanes as part of a natural cycle. Some researchers at NOAA and Colorado State University have conducted studies that support the theory that the number and intensities of hurricanes follow in 50- to 70-year cycles called the Atlantic *Multidecadal* Oscillation (AMO). They believe the AMO is controlled by gradual changes in the North Atlantic Ocean currents.

The AMO controls the flow and direction of the major wind systems. When the *trade winds* blow steadily from the east, they produce excellent conditions for hurricanes to form. Right now Earth is in the AMO, which is why there are presently so many hurricanes. When the strength of the ocean currents changes and causes the *westerlies* to move southward toward the trade winds, this keeps hurricanes from forming. During these times of the cycle, there are not many hurricanes.

It has also been suggested that El Niño events in the Pacific, which occur every four to seven years, tend to keep hurricanes from forming in the Atlantic, especially strong hurricanes of category 3 or higher. El Niños occurred in the Pacific in 1997 and 2006. These two years also had very little storm activity in the Atlantic.

The 2004–05 hurricane season spurred much debate among climate scientists. Kevin Trenberth at the National Center for Atmospheric Research located in Boulder, Colorado, believes that half of 2005's extra ocean warmth was due to global warming. Clifford Jacobs at the National

Science Foundation (NSF) calls the situation a "raging scientific debate." According to the National Research Council, the past few decades have been the warmest on Earth in the last 400 years, if not the last several thousand years. One of the panel members of the National Research Council, Kurt Cuffey, believes there is an overwhelming amount of evidence that human activities—such as deforestation, farming practices, and the burning of fossil fuels—have significantly contributed to global warming and thus to the number of hurricanes per year.

One of the critical aspects of hurricanes is the amount of destruction they cause. The harm to people and damage to property can be enormous if they strike heavily populated areas. Unfortunately, coastal areas are some of the most desirable places to live, as evidenced by some of the United States's most populated cities: New York, Los Angeles, Miami, and New Orleans.

The remains of the middle school in Pass Christian, Mississippi, September 19, 2005. Hurricane Katrina caused extensive damage all along the Gulf Coast. *(Mark Wolfe, FEMA)*

Biloxi, Mississippi, April 1, 2006. After Hurricane Katrina, demolition was the only choice for many buildings such as this one along Highway 90. After seven months, it was still difficult to comprehend the degree of devastation the Mississippi coast area sustained. *(George Armstrong, FEMA)*

The period from 1970 to 1994 was a relatively quiet period in terms of tropical cyclone activity. In fact, relatively few actually made landfall anywhere in North America. During this time, the United States's population was beginning to grow, and expansion of settlements along the nation's coasts was enormous. Similar tendencies of population growth along coasts has occurred worldwide.

When major hurricanes hit populated areas the effects can be disastrous. Even people that have "hurricane proofed" their homes with cement reinforcement and hurricane shutters have returned after a storm to find their house completely gone. Some residents who lose their homes after major hurricanes never rebuild but move from the area instead. (This was the case after Katrina, which hit the U.S. South-

Pass Christian, Mississippi, October 4, 2005. An aerial photo of destroyed Mississippi Gulf Coast Highway I-90 as a result of winds and tidal surge from Hurricane Katrina. This section of the bridge connected Pass Christian, near Gulfport, to Bay St. Louis. *(John Fleck, FEMA)*

Punta Gorda, Florida, in 2004 after Hurricane Charley swept through
(Andrea Booher, FEMA)

east in 2005.) Others, who return, face major reconstruction. Often those who do not lose their house have to deal with the eventual invasion of indoor mold due to the moisture from the storm, causing them to evacuate later. Whether storm victims relocate, rebuild, or choose some other alternative, the impact has a financial, emotional, logistical, and psychological cost. Family members are often lost, as well as property and other personal items.

When people rebuild, not only is it very expensive, but families must also find temporary places to live in the meantime. New construction can take months, or even years, to complete. Some people have had to find nearly three years of temporary housing before they have a permanent home to live in once again.

It is important to understand, too, that just because most recent hurricane activity has occurred in the southeastern portion of the United States in the past few years, hurricanes can also travel up the eastern seaboard to areas such as New York, as well as west to Texas, for example especially with the increasing effects of global warming and the changing character of recent weather patterns.

HURRICANES OF THE PAST

This list highlights some of the most notable hurricanes to strike the United States.

1893—New York. In August 1893, a category 2 hurricane made landfall in the area where the John F. Kennedy International Airport is located today. It swept through Manhattan's Central Park and uprooted hundreds of trees. A 30-foot (9-m) storm surge traveled through Brooklyn and Queens and destroyed almost every human-made structure in its path. And then there was an island named Hog Island, a pig-shaped mile-long barrier island off the southern coast of the Rockaways. When the hurricane passed over it, it disappeared. It is the only recorded incidence of the removal of an entire island by a hurricane.

1938—The Long Island Express. The hurricane called the Long Island Express by many is one of the most unforgettable hurricanes in U.S. history. Racing across the Atlantic, it originally headed for Florida, but then shifted north and traveled up the east coast. The storm was moving so fast—greater than 60 miles (97 km) per hour, in other words, almost twice as fast as normal—that it was impossible to track and predict its movement.

When the hurricane and its tidal surge hit Long Island, New York, the impact was so strong it registered on seismographs thousands of miles away in Alaska. The storm was enormous: Its eye alone measured 50 miles (80 km) wide. The hurricane continued traveling 50 miles (80 km) per hour northward into the New England states; wind gusts measured 120 miles (193 km) per hour. One area in Rhode Island was inundated with surge water that started out just a few inches deep, then was waist

HURRICANE KATRINA AND GLOBAL WARMING

Hurricane Katrina struck Louisiana and Mississippi and other southeastern states in 2005 and wreaked so much havoc and destruction on those areas of the United States that many people have been looking at global warming and wondering if that was the cause.

deep within minutes, and continued rising until it was more than 13.5 feet (4 m) deep.

An *anemometer* in Massachusetts recorded sustained wind speeds of 121 miles (195 km) per hour. This reading holds the record today as the second-highest winds ever recorded on Earth. Eventually, the storm moved into Canada and the Arctic, where it died out. More than 690 people were killed in this single storm.

1972—Hurricane Agnes. According to NOAA, Hurricane Agnes set inland flood records across the Northeast. Its damage was estimated at $3.2 billion. Agnes came out of the Gulf of Mexico and traveled from Florida up to New York. More than 210,000 people were forced to evacuate their homes, and 122 people were killed. Agnes was unusual because it flooded so far inland. It impacted areas several hundreds of miles from the coasts. The entire state of Pennsylvania was declared a disaster area. In the central part of Virginia, almost every creek and stream overflowed its banks. Florida experienced high tides, winds, and even tornadoes.

Because of severe weather incidents like these, the National Weather Service's modernization efforts in cutting-edge technology, such as Doppler radar, weather satellites, and automated river and rain gauges, are able to assist in monitoring severe weather. Another forward-looking improvement is in emergency preparedness. Forecasters are now able to broadcast weather information to the public and emergency facilities faster as communications technology continues to improve. Public awareness has also improved, and more families and businesses have implemented emergency evacuation plans.

Source: NOAA

Unfortunately, science cannot tell climatologists directly if global warming specifically causes hurricanes. Scientists cannot positively say, therefore, that global warming directly caused Katrina or any other specific incident of occurrence. But just because Katrina cannot be directly linked to global warming as the sole, specific cause, it does not mean that global warming did not play a significant role in Katrina's occurrence.

Climatologists at NOAA can say with certainty that they understand the physical mechanisms of how hurricanes are formed, what sustains them, how they grow, and how they move. Because they know for a fact that hurricanes draw their strength from warm ocean water, the warmer the water is, the more powerful the storm can be. Records of sea-surface temperature confirm that the ocean water is currently more than 1°F (0.6°C) warmer than it was a century ago. This temperature can fluctuate from year to year, too, if the water gets warmer, it increases the chance of supporting a hurricane during hurricane season. It is suggested that this happened with Katrina. While Katrina was strengthening from a tropical storm to a category 5 hurricane, as it traveled between the Florida Keys and the Gulf Coast, the surface waters in the Gulf of Mexico were unusually warm—2°F (1.2°C) warmer than normal.

While there is no way to prove global warming caused Katrina, it is reasonable to say that it increased the probability of Katrina developing. Some scientists believe that global warming has created an environment under which powerful storms such as hurricanes are more likely to occur now and in the future. For example, a study published in *Nature* notes that during the second half of the 20th century, the average intensity of tropical storms has increased globally. This matches the same time period that sea surface temperatures have been rising. Because warm oceans are such a critical ingredient in hurricane formation, as the ocean temperatures rise, the probability of hurricanes will likely rise as well.

Not all scientists agree on this issue, however. There is some controversy in the science arena on the connection between hurricanes and global warming. Christopher Landsea, a scientist at the National Hurricane Center in Miami, Florida, believes that it only appears there are more hurricanes recently because they are monitored better than they were 100 years ago. He contends that back in the early 1900s, the only reports gathered were by ships crossing the ocean. Today there is complete satellite coverage of Earth. Landsea believes, instead, that there are multidecadal swings of "active" then "quiet" periods for the storms. Even so, Landsea does not say that global warming has not had an impact on hurricanes—just that he does not know if it has had a big enough influence to say specifically that global warming has contributed a measurable influence on hurricane frequency and intensity.

Other researchers debate past records kept in the late 1800s and early 1900s, many claiming the records inaccurately represent the true number of storms because the only method of collecting data was via ships (satellite surveillance systems were not available yet). NOAA's National Hurricane Center, which maintains HURDAT (the official record of tropical storms and hurricanes for the Atlantic Ocean, Gulf of Mexico, and Caribbean Sea) each year adds data from the hurricane season. Major revisions and reanalyses were made to the database in 2003 to update the time period from 1850 through the early 1900s.

Concerning the update and reanalysis, Chris Landsea remarked, "There are many reasons why a reanalysis of the HURDAT dataset is both needed and timely. HURDAT contained many systematic and random errors that needed correction. Additionally, as our understanding of tropical cyclones had developed, analysis techniques at the NOAA National Hurricane Center changed over the years and led to biases in the historical database that had not been addressed. Another difficulty in applying the hurricane database to studies concerned with landfalling events was the lack of exact locations, time and intensity information at landfall.

"Finally, recent efforts led by the late José Fernandez-Partagas, a Cuban research meteorologist in Miami, Florida, uncovered previously undocumented historical tropical cyclones in the mid-1800s to early 1900s that have greatly increased our knowledge of these past events, which also had not been incorporated into the HURDAT database."

More than 5,000 additions and alterations were approved for the years 1851 to 1910 by the NOAA National Hurricane Center's Best Track Change Committee.

According to Michael E. Mann, associate professor of meteorology at Penn State, and director of the Earth System Science Center, many scientists believe that the numbers of historic tropical storms in the Atlantic are seriously undercounted. A statistical model he developed, however, shows that the estimates currently used are only slightly below modeled numbers and indicate that the numbers of tropical storms in the recent past are increasing.

"We are not the first to come up with an estimate of the number of undercounted storms," Mann said. He believes that in the past, researchers assumed that a constant percentage of all the storms made landfall, and so they compared the number of tropical storms making landfall with the total number of reported storms for that year. Other researchers looked strictly at ship logs and ship tracks to determine how likely it was a tropical storm would have been missed. In the early 1900s and before, there were probably not sufficient ships crossing the Atlantic to obtain full coverage.

Researchers reported that "the long-term record of historical Atlantic tropical cyclone counts is likely largely reliable, with an average undercount bias at most of approximately one tropical storm per year back to 1870." This contrasts with the prior estimate of three or more undercounted storms.

Mann's model looked at how the El Niño/La Niña cycle, the pattern of the Northern Hemisphere jet stream and tropical Atlantic sea-surface temperatures, influences tropical storm generation by creating a model that includes these three climate variables. The input information was available back to 1870.

The model, trained on the tropical storm occurrence information from 1944 to 2006, showed an undercount before 1944 of 1.2 storms per year.

In addition, according to Kerry Emanuel, a professor of meteorology at the Massachusetts Institute of Technology, Landsea does not acknowledge that there was also an eastward shift in sea surface temperatures, which caused more hurricanes to form farther to the east and fewer to strike the U.S. coastlines in recent years—so that even though a hurricane may not have made landfall on the U.S. coast, it does not mean one did not occur. Another important factor is that hurricanes are currently forming closer to the equator, which gives them more warm water to move across and pick up additional energy so that they become more intense, destructive storms.

Scientists also link hurricane frequency with El Niño patterns. When El Niño occurs, there are fewer hurricanes. Scientists believe

THE COSTLIEST U.S. HURRICANES

The following table illustrates the 30 most costly tropical cyclones to strike the U.S. mainland, according to the National Weather Service's National Hurricane Center. (Damages are listed in U.S. dollars, inflation adjusted to the year 2004, with the exception of Katrina 2005.)

RANK	HURRICANE	YEAR	CATE-GORY	DAMAGE
1	Katrina (Florida, Louisiana, Mississippi, Alabama)	2005	5	$81,200,000,000
2	Andrew (southeast Florida, southeast Louisiana)	1992	5	$43,672,000,000
3	Charley (southwest Florida)	2004	4	$15,000,000,000
4	Ivan (Alabama, northwest Florida)	2004	3	$14,200,000,000
5	Hugo (South Carolina)	1989	4	$12,250,000,000
6	Agnes (Florida, U.S. Northeast)	1972	1	$11,290,000,000
7	Betsy (southeast Florida, southeast Louisiana)	1965	3	$10,799,500,000
8	Frances (Florida)	2004	2	$8,900,000,000
9	Camille (Mississippi, southeast Louisiana, Virginia)	1969	5	$8,889,000,000
10	Diane (U.S. Northeast)	1955	1	$6,997,700,000

(continues)

(continued)

11	Jeanne (Florida)	2004	3	$6,900,000,000
12	Frederic (Alabama, Mississippi)	1979	3	$6,291,000,000
13	Great New England Hurricane, or Long Island Express (New York, New England states)	1938	3	$5,971,000,000
14	Allison (north Texas)	2001	Tropical storm	$5,829,000,000
15	Floyd (Mid-Atlantic states, U.S. Northeast)	1999	2	$5,764,000,000
16	Great Atlantic Hurricane (U.S. Northeast)	1944	3	$5,386,000,000
17	Fran (North Carolina)	1996	3	$4,525,000,000
18	Alicia (north Texas)	1983	3	$4,384,000,000
19	Opal (northwest Florida, Alabama)	1995	3	$4,324,000,000
20	Carol (U.S. Northeast)	1954	3	$3,949,000,000
21	Isabel (Mid-Atlantic states)	2003	2	$3,643,000,000
22	Juan (Louisiana)	1985	1	$3,105,000,000
23	Donna (Florida/ eastern United States)	1960	4	$3,040,000,000
24	Celia (south Texas)	1970	3	$2,761,000,000
25	Bob (North Carolina, U.S. Northeast)	1991	2	$2,593,000,000

26	Elena (Mississippi, Alabama, northwest Florida)	1985	3	$2,588,000,000
27	Carla (north and central Texas)	1961	4	$2,366,000,000
28	Great Miami Hurriane (Florida, Mississippi, Alabama)	1926	4	$2,058,000,000
29	Eloise (northwest Florida)	1975	3	$2,008,000,000
30	1915 Galveston Hurricane (north Texas)	1915	2	$1,990,000,000

Source: NOAA

that compared to the 2005 hurricane season, 2006 was relatively quiet because it was an El Niño year. Scientists do not currently know, however, how global warming and climate change will affect El Niño.

Scientists acknowledge that with global warming, the resultant warmer sea surface temperatures and extra heat in the ocean serves as a means to generate destructive tropical storms. Experts also agree that as more people move to the coastal areas and continue to build and develop in vulnerable zones, there will continue to be disasters like Katrina in 2005, causing loss of life and extensive damage to property.

HURRICANE RESEARCH

NOAA is heavily involved in hurricane research and is approaching the issue from several angles, such as conducting theoretical studies, developing sophisticated computer models, and collecting measurements in the field from actual hurricanes that can be analyzed. Because hurricanes are such complex storms, NOAA's goal is to improve scientific understanding of how and why they form; why they behave as they do; what mechanisms cause them to strengthen and weaken; how they can be predicted; how better warning, emergency, and evacuation plans can be put in place; and how lives and property can best be protected.

In the field, NOAA's Atlantic Oceanographic and Meteorological Laboratory (AOML) uses remotely-sensed observations of both hurricanes and the ocean underneath their predicted path. The lab also makes direct observations inside the inner core of the hurricane. The AOML uses many types of sophisticated instruments, such as radars, fixed probes, and expendable probes to measure wind speed, temperature, and pressure, as well as wave height and sea surface temperature. These measurements allow NOAA to predict hurricane intensity, track hurricane paths, and predict damage impact.

NOAA is also heavily involved in computer-modeling research of hurricanes. Its Geophysical Fluid Dynamics Laboratory uses models to help NOAA scientists understand a hurricane's growth, development, and decay and its relationships to atmospheric and oceanic processes. The NOAA Environmental Modeling Center is currently working on the development of the next generation of hurricane forecast modeling systems. Eventually NOAA's National Hurricane Center will use this model to create real-time tropical forecasts by its hurricane specialists.

TORNADO AND THUNDERSTORM RESEARCH

Some believe one of the effects of global warming is an increase in extreme, violent weather events in the form of tornadoes and thunderstorms. In 2004, there were more tornadoes reported in the United States than since records have been kept. In Kansas alone, there were 124 reported tornadoes. According to NOAA's Storm Prediction Center in Norman, Oklahoma, nationwide, there were 1,555 tornadoes recorded from January through September 2004. This broke the previous record, set in 1998, by more than 130 tornadoes. One positive development in tornado research is that improvements in technology now allow meteorologists to spot more rapidly rotational patterns in thunderstorms and issue tornado warnings to the public before a destructive funnel forms.

THE TEN WORST PLACES FOR AN EXTREME HURRICANE TO STRIKE

As coastal city populations continue to grow, they face the threat of greater losses to life and property. The following locations in the United States would incur the heaviest losses if an extreme hurricane were to make landfall.

RANK	LOCATION	POSSIBLE INSURED LOSSES	POTENTIAL ECONOMIC LOSSES
1	Miami/Ft. Lauderdale, Florida	$61.3 billion	$122.6 billion
2	New York City, New York	$26.5 billion	$53 billion
3	Tampa/St. Petersburg, Florida	$25.1 billion	$50 billion
4	Houston/Galveston, Texas	$16.8 billion	$33.6 billion
5	New Orleans, Louisiana	$8.4 billion	$16.8 billion
6	Mobile, Alabama	$6.0 billion	$12 billion
7	Boston, Massachussetts	$5.1 billion	$10.2 billion
8	Biloxi/Gulfport, Mississippi	$5.1 billion	$10.2 billion
9	Myrtle Beach, South Carolina	$4.3 billion	$8.6 billion
10	Norfolk, Virginia	$3.9 billion	$7.8 billion

Source: The Consumer Insurance Guide

This is an occluded mesocyclone tornado. *Occluded* means "old circulation" on a storm; this tornado was forming while the new circulation was beginning to form the tornadoes that preceded the F5 Oklahoma City tornado. This location is seven miles (11 km) south of Anadarko, Oklahoma. *(NOAA, National Severe Storms Laboratory Collection)*

The National Severe Storms Laboratory (NSSL) is presently testing a new type of radar for weather detection called "phased array" that is able to scan the sky six times faster than the current Doppler radars in use. The NSSL also recently made a modification to Doppler radar called "dual polarization" that will provide meteorologists with more detail about what is going on inside storm cells. The lab is also using low-frequency sound waves, or "infrasound," to try to predict and detect tornadoes. Meteorologists also use computers to model severe weather in order to understand better violent storms and how to predict them.

NOAA is interested in improving its scientific understanding and predicting abilities of tornadoes and thunderstorms so it can more effectively monitor, track, and warn the public of emergencies. This in turn will hopefully minimize loss of life and damage to property.

Meanwhile, other scientists are researching ways to prevent tornadoes from forming to begin with. A project has been developed where using a burst of microwave energy specifically beamed down from a carefully calculated direction from a space satellite could be used to destroy the destructive power of a tornado cell in a thunderstorm. According to the American Meteorological Society, roughly 1,200 tornadoes on average are reported each year in the United States. These destructive storms average 55 fatalities annually and cause billions of dollars worth of property damage. Because climatologists have begun to warn that an increase in global warming may cause an increase in tornadoes in the future, scientists have been experimenting in the development of a device to negate their formation and destructive effects. A revolutionary new concept was presented at the American Society of Civil Engineers' Space 2000 Conference and Exposition on Engineering, Construction, Operations and Business in Space. Called a "thunderstorm solar power satellite," this revolutionary new concept proposes a system that beams microwave energy into the downdraft region of a thunderstorm, the portion in the system where the funnel cloud initially forms. Bernard Eastlund of Eastlund Scientific Enterprises Corp. and Lyle Jenkins of Jenkins Enterprises, who developed the system, believe that a pulse of microwave energy would serve to disrupt the convective flow that the tornado needs to concentrate the energy and form the funnel. By using these satellites as a type of high-tech "space-age gun," they believe the bursts of extremely well-placed strikes of microwave energy could change the temperature and structure of storm systems. According to Jenkins, "We call it taming the tornado. With just a little burst of microwave energy, we think we see a way to negate the trigger point in tornado creation. We want to heat the cold rain. By tailoring the beam, it could absorb the rain that is part of the tornado-making process."

In essence, heating the rain breaks up the downdraft that gives the tornado its energy. The team envisions placing their "tornado-stopping satellites" in a geosynchronous orbit, where the satellite is always positioned over the same spot on Earth. In this case, the satellites would be positioned over the areas in the United States that are commonly subject to tornadoes, such as Oklahoma, Kansas, and Texas. They also envision adding the function of Doppler radar in order to detect the initial formation processes of tornadoes, giving them greater advanced warning. The scientists involved in this research acknowledge that more investigation is needed to determine how much microwave energy the bursts should have to be the most effective. Researchers are hopeful that in this new field of weather modification, the success of modifying tornadoes could also be applied to hurricanes in the future.

Climate Research—
What the Experts Say

Climate research is a rapidly expanding field. The advent of more sophisticated technologies such as satellite monitoring systems and high-speed computers has enabled scientists to increase the efficiency of field research, collect and process more data, and be able to put meaning to large amounts of technical data that cross technical and professional boundaries from one science specialty to another, such as paleoclimatology, oceanography, meteorology, physics, chemistry, geography, geology, and a host of other physical sciences. This chapter presents some of the most recent technological research being carried out in the field at present.

One of the most challenging aspects of acquiring a detailed understanding of the world's oceans and how they interact not only with land and the atmosphere but with each other is the lack of data over the vast expanses of ocean. Being able to obtain water samples and take salinity and temperature measurements in a systematic way across the entire planet has not been feasible technically, economically, and logistically, making it extremely challenging to create a robust database over time. With the advent of satellite technology, however, it finally became technically and logistically feasible to begin collecting reliable global sea-surface temperatures. Another advantage of satellite technology is that it is possible to remove the atmospheric effects of clouds and aerosols in the atmosphere from the data, maintaining the data's integrity. Satellites can provide sea surface temperatures that are accurate to within 0.8°F (0.5°C). Data on wind speed and direction is also collected. Data collection from satellites allows scientists to understand better the interactions between the oceans and atmosphere. Satellite data obtained by NASA is acquired from the *Terra* spacecraft and the *QuikSCAT* (a microwave radar). Through these instruments, NASA is better able to understand the processes of ocean circulation. In addition to these instruments, scientists also use other satellites, such as *Landsat 7*. Data from these satellites allows scientists to build and test general circulation models, as well as model the ocean's physical, biological, and chemical inputs toward global climate change.

Several areas of ongoing research today are lending clues to global warming and the response of Earth to these changes. Like putting the pieces of a puzzle together, each additional piece of information that scientists collect enables them to refine the bigger picture of global warming.

TRACKING THE OCEAN'S CIRCULATION

To understand climate change, it is necessary to be able to look back in time and determine what Earth's past circulation patterns were like. Once this has been established, those patterns can be compared to today's, increasing knowledge about this complex system. This, in turn, gives climatologists the information they need to develop computer models to predict future ocean circulation. One of the biggest questions scientists have asked is whether there is some sort of a "switch" that can be turned on or off and can speed up, slow down, or stop the North Atlantic conveyor belt current. The conveyor belt represents a huge force in the ocean. In fact, according to Jerry McManus and Delia Oppo at the Woods Hole Oceanographic Institution, this conveyor belt

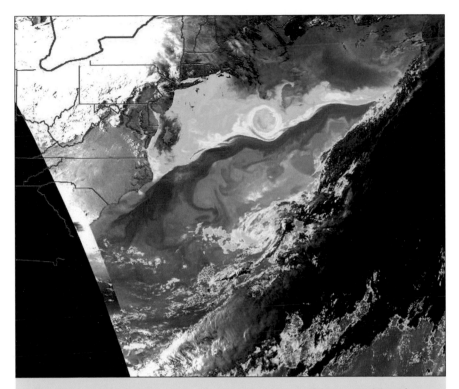

This thermal image illustrates how scientists are able to study and track the ocean's surface temperatures. This image, obtained from the *AVHRR* satellite, records the warmest temperatures in red, orange, and yellow and the coolest temperatures in blues and green. The warm area in this image depicts the location of the Gulf Stream. *(Johns Hopkins University Applied Physics Laboratory and Ocean Remote Sensing Group)*

responsible for overturning the ocean is equal to roughly 20 times the combined flow of all the world's rivers.

Because of the tremendous amount of heat and energy this current carries, it is critical to understand how and why it behaves as it does. In this attempt, several computer models have been developed. In order to make the models as reliable and realistic as possible, scientists turn to paleoceanography to understand past currents and energy flow and their impacts on climate.

Scientists are able to look back in time by studying proxy data found in sediment cores and dating the fossilized shells of foraminifera.

The analysis of foraminifera can designate when specific water masses formed. In addition, the radioactive decay of naturally-occurring uranium in seawater to its daughter isotopes (protactinium and thorium) found in deep-sea mud is also used to determine how strong and fast the water currents moved.

Of the two daughter isotopes, protactinium stays in the seawater for centuries. This is important because it lasts long enough to be transported to the Southern Ocean by the giant circulation cell. When the conveyor belt slows, the proportion of protactinium buried in the North Atlantic sediments increases. Based on this fact, scientists can use the ratio of protactinium-to-thorium levels in the sediments to see how fast the overturn circulation was moving, thereby providing important clues to the past climate.

For the future, if global warming is not checked and slowed, it could cause the addition of freshwater to the ocean, disrupting the system. An increase in temperature could increase evaporation at lower latitudes and transport freshwater toward the poles where it will return to Earth as snow or into the ocean. Based on all of these studies, it is apparent that there is still a lot of work that needs to be done in order to understand better the synergy of the biosphere, atmosphere, and hydrosphere.

CO_2 SEQUESTRATION

One of the key issues in global warming and one in which the ocean plays a significant role, as illustrated earlier, is the storage of CO_2. Long-term sequestration of carbon dioxide in the ocean waters directly affects the amount of CO_2 in the atmosphere by providing a reservoir that balances out the overloading of CO_2 into the atmosphere by humans. The key research questions here are how much CO_2 the oceans can effectively store, how long it takes to store it, whether the storage is permanent, and what effect the CO_2 may have on marine ecosystems.

According to a study by Adina Paytan, an oceanographer at the University of California at Santa Cruz, the last time Earth experienced an extreme global warming event, the oceans acted as a buffer and helped slow the warming process. Her research indicates that during the period of Earth's history referred to as the Paleocene-Eocene thermal maximum (PETM), 55 million years ago, ocean plant life productivity increased and served as a mechanism to capture excessive carbon from the atmosphere. The plant life eventually settled to the ocean floor where it was buried—effectively sequestering the CO_2. Paytan stresses that the mechanism was increased productivity and not just changes in ocean chemistry, as others had proposed.

According to Paytan, when the plant matter settles at the bottom of the ocean, it contains barite (barium sulfate) in its tissues. When the barium reaches its saturation point on the ocean floor, it combines with sulfur and forms the mineral barite. Paytan believes it may have been this sequestration process that eventually brought down the warmer Earth temperatures and suggests that the same process may play an important role today in Earth's adjustment to global warming. The major drawback, she cautions, is that it is a slow, natural process—much slower than the current human-caused buildup of carbon occurring today. She determined that the barite accumulation in the PETM took 170,000 years, illustrating that it is a lot easier to heat up Earth than to cool it down again.

Another theory surrounding the events of the PETM has been proposed by the scientists at NASA's GISS. According to Gavin Schmidt there, large quantities of CO_2 were stored in a reservoir of methane hydrate deposits buried on the continental shelves of the oceans. Methane hydrates are formed when bacteria decompose organic matter. During the process, bacteria produce methane, and in cold, high-pressure environments, methane hydrates are formed.

Another study, conducted by British scientists that appeared in the November 2007 edition of the *Journal of Geophysical Research,* has suggested that the world's oceans—which scientists believe is a tremendous reservoir for the CO_2 being added to the atmosphere—are beginning to lose their ability to absorb CO_2 from the atmosphere. According to data they collected, the North Atlantic's absorption of CO_2 dropped in half between 1995 and 2005.

These researchers warn that if the ocean loses its ability to absorb CO_2, it will increase the effects of global warming felt on land. Through their research it was also determined that the oceans were becoming more acidic because of the CO_2 uptake in the ocean. This presents a compound problem: If the oceans stop absorbing as much CO_2, then the negative effects of global warming will increase; if the oceans absorb more CO_2, the oceans become more acidic, which is harmful for marine ecosystems and *biodiversity*.

Some scientists warn to be cautious about making assumptions concerning findings, stating it is still too early to make concrete conclusions. How much is natural variation and how much is anthropogenic still needs to be determined. What is stressed, however, is that the ocean's function as a CO_2 "sink" can change quickly and needs to be monitored closely for subtle changes that may offer clues to long-term climate change and global warming.

In another study, whose results were published in the *Proceedings of the National Academy of Sciences,* it was determined that economic growth had caused levels of atmospheric CO_2 to increase 35 percent faster since 2000 than had been expected. Pep Canadell of the Global Carbon Project attributed this increase to an 18 percent decline in the efficiency of CO_2 sinks (the world's oceans and forests), laying the rest of the blame on the increased use of fossil fuels.

LARGE-SCALE SALINITY CHANGES

According to the Woods Hole Oceanographic Institution, tropical ocean waters have become much saltier over the past 40 years, while the polar waters have become much fresher. In addition to this, there is increased warming at Earth's surface, which has increased evaporation over the low-latitude (equatorial) oceans. When water is evaporated, it picks up freshwater, leaving the salt behind. Carried by Earth's circulation patterns, the freshwater moves toward the polar regions where it is deposited as precipitation, adding still more freshwater to those extreme latitudes.

Global warming may be one of the processes that is contributing to this shift. A study conducted by Ruth Curry (at Woods Hole), Bob Dickson (of the Centre for Environment, Fisheries, and Aquaculture Sciences, in the United Kingdom), and Igor Yashayaev (at the Bedford Institute of Oceanography, in Nova Scotia, Canada) suggests that an acceleration of Earth's global water cycle can cause far-reaching effects, leading to droughts, floods, and storms. In addition to the current melting of polar glaciers, the influence of global warming enhances the process by rapidly adding more water, which causes the polar regions to become fresher. The long-term concern is the disruption of the Atlantic Ocean conveyor; if this major ocean circulation is altered, it will trigger other climate changes (see chapter 6 for a more in-depth discussion).

During their study, the researchers collected ocean water samples over a broad area—from Greenland to the southern tip of South America. They concluded that water masses in the Atlantic have been actively changing to this equatorial-salty/polar-fresh pattern over the past 50 years. Since 1990, the process has accelerated, corresponding to the same time interval of the 10 warmest years of recorded temperatures on Earth. Net evaporation in the Tropics is currently estimated at 5 to 10 percent, leading scientists to conclude that the loss of freshwater from the Tropics and the subsequent increase of freshwater to the polar regions are happening faster than the natural circulation of the ocean can keep up with. According to GISS, based on data collected from the Mediterranean Sea and Pacific and Indian Oceans, similar trends are occurring there as well, suggesting that the water cycle is being affected globally. These conditions could cause significant climate changes to occur on timescales of just decades, not centuries or longer.

LOCATING 64°F WATER

According to Terry Joyce at Woods Hole, there exists a phenomenon in the North Atlantic Ocean of distinct parcels of water with constant salinity, density, and temperature; these pockets maintain a temperature of 64°F (18°C). Initially discovered by Valentine Worthington in the Sargasso Sea in the early 1970s, he believed these specific water masses were connected in some important way to the circulation of the entire North Atlantic and the weather associated with it.

Today, researchers have discovered that a specific energy transfer between warm Gulf Stream water and the cold winter atmosphere produce this water. In an attempt to understand how this water influences the North Atlantic climate, a program called the CLIVAR Mode Water Dynamics Experiment (CLIMODE) was begun in November 2005.

This layer of water forms during the winter when frigid winds blow from the Arctic out to where the much warmer Gulf Stream flows in the Sargasso Sea. The cool winds pick up the heat from the Gulf Stream and deliver it to Europe, keeping their climate warmer. The cold, salty water that is left behind forms a surface layer that can extend down to 1,300 feet (400 m). When the ocean warms during the spring and summer, this cold layer sinks to the middle level of the ocean and stays suspended between the warm surface waters and the extremely cold deep waters.

Because of the unusual ocean layer's persistence, Joyce asserts this water acts as a "memory bank" of North Atlantic climate; it basically freezes a memory of conditions from the winter that it was formed in and carries it around the ocean. Oceanographers continually collect samples of this water for future analysis, hoping they will be able to determine how fast or slow the ocean-atmosphere system responds to climate change and global warming. Researchers also want to understand better these unique water layers. They want to know how they affect general ocean circulation; what role they play in the storage of CO_2, nutrients, and heat from the ocean's surface; and how they affect marine life.

ATLANTIC FRESHWATER

According to Curry at Woods Hole, several large areas of the North Atlantic have been getting fresher since the 1960s due to the effects of glacial meltwater flowing into the ocean and increased precipitation from an enhanced water cycle. Both Curry and Cecilie Mauritzen of the Norwegian Meteorological Institute have mathematically determined how much additional freshwater has been added to cause the change, how fast it entered ocean circulation, and where the freshwater was stored.

Based on the past 40 years of data, they believe that the continued freshening of the polar waters could negatively affect ocean circulation (such as the Atlantic conveyor) within a century, or even faster considering the increased effects of global warming. Curry and Mauritzen determined that in an average year, 1,200 cubic miles (5,000 km³) of freshwater flows from the Arctic into the North Atlantic near Greenland. In addition, 4,558 cubic miles (19,000 km³) flowed into the northern seas from 1965 to 1995. As a comparison, the outflow of the Mississippi River each year is about 120 cubic miles (500 km³), and the Amazon's is 1,200–1,439 cubic miles (5,000–6,000 km³).

In addition, between 1970 and 1995, the polar areas experienced an additional freshening with an influx of freshwater. In an episode called the "Great Salinity Anomaly," a massive amount of freshwater entered the Nordic seas in the late 1960s and traveled toward the south in the East Greenland Current. This episode alone added 2,399 cubic miles (10,000 km^3) of freshwater to the northern oceans.

According to Curry, another significant influx of freshwater from the Arctic entered the ocean again in the 1980s and 1990s, concentrating in the polar areas. This freshwater "pooling" is important because it interferes with the great ocean conveyor belt in the North Atlantic that is responsible for distributing heat from the Tropics to the poles. These huge influxes of freshwater have the capability to slow down or even stop the circulation. Curry reports that so far, the freshening does not appear to have had a detrimental impact, but it is expected that this freshening will continue over the next few centuries, and continued freshening of the ocean waters in the polar regions could cause the conveyor belt to slow, creating problems worldwide. Another factor to consider is the enormous amount of freshwater contained in Greenland's ice sheet. If global warming persists and the ice sheet continues to melt, this represents another significant source of freshwater influx into the polar oceans. Curry estimates that at the rate of influx seen over the last 30 years, it would take about a century to add enough freshwater—about 2,159 cubic miles (9,000 km^3)—in the Nordic seas to slow the ocean circulation and roughly two centuries of continued dilution to stop it completely.

She also cautions there are uncertainties involved, such as future rates of greenhouse warming and glacial melting. Based on the most recent computer models of the situation, greenhouse warming predicts an increase in precipitation and runoff at high latitudes that leads to a slowdown of the conveyor belt current. The function of glaciers is important because they are considered to be the "wild card" in this scenario; the addition of glacial meltwater, the collapse of ice shelves into the ocean, and surges of glacial movement could all accelerate the process of adding freshwater to the ocean.

SOUTHERN OCEAN CURRENTS

It is not just currents in the Northern Hemisphere that are affected by global warming. Those in the Southern Hemisphere are faced with the same issues. Around Australia, the southern oceans are also being affected in the same way that those in the north are. As the glaciers and ice sheets melt in Antarctica, they are adding significant amounts of freshwater, which have a negative impact on the formation of the ocean's dense bottom layer of water. This layer is responsible for driving the ocean's global circulation system.

According to Steve Rintoul, a senior scientist at Australia's CSIRO Marine Science, if enough freshwater is added to the ocean from Antarctica, it will keep it from being salty enough to sink and flow northward, where it transports heat worldwide. Although the incident would be local, Rintoul says, the effects would be felt around the globe. He attributes greenhouse gas and global warming to the problem. In addition, the Australian scientist's fieldwork revealed that the Southern Ocean waters are becoming more acidic because they are absorbing much of the CO_2 emitted from polluting, industrialized countries. This presents a serious situation because it impacts the ability of plankton to absorb CO_2 and store it at the bottom of the ocean in a carbon sink on the ocean floor, as it was previously doing.

A third aspect Rintoul stresses is that global warming is also impacting the wind patterns in the Antarctic and causing them to change. Wind systems are migrating to the south, away from Australia, which is causing drought conditions in the western coastal areas. This, in turn, is negatively impacting Australia's agricultural resources through reduced productivity.

TRADE WIND CHANGES

A study conducted by Gabriel Vecchi of the University Corporation for Atmospheric Research in Boulder has suggested that the trade winds of the Pacific Ocean are weakening because of global warming. There is a large wind system that circulates over the Pacific Ocean called the "Walker circulation." Through the use of data collected in the field, as well as computer modeling, Vecchi has determined that the Walker circulation has weakened by roughly 3.5 percent since the 1850s. It is expected that by 2100, it will have weakened another 10 percent.

Vecchi concludes that human-induced climate change and global warming are playing a large role in the situation. He emphasizes that altering the current is significant because it could disrupt food chains, which would then negatively impact the biological diversity of the area.

The team of scientists involved in the study used two different types of models: 1) those that factored in the influence of greenhouse gases contributed by human activity, such as the burning of fossil fuels, and 2) those that looked only at natural forcings that affect climate, such as solar variations and volcanic activity. Vecchi concluded that the only way the models could realistically explain the data they had collected was with the model that included human activity and the addition of greenhouse gases from global warming.

Based upon the data they collected and the results of the models, Vecchi and his team of researchers concluded that the Walker circulation could very well slow another 10 percent by the end of this century. The slowing of this circulation pattern would have several negative effects on Earth's climate because it plays a significant role in moving ocean currents. It directly affects the upwelling process that occurs on the coast of South America, providing rich fishing waters for the economies of that region. It also provides nourishment for marine life all along the equatorial Pacific.

Vecchi explains the circulation system mechanism as follows. In order to maintain a proper energy balance, rates of evaporation must equal rates of precipitation. This balance gets upset when temperatures rise, because it causes more evaporation of water into the atmosphere. Even though there is much more moisture in the lower atmosphere, it does not fall as rain. This creates an imbalance in the system because the atmosphere begins gaining water faster than it can release it, which causes the air circulation to slow down to keep the system "balanced." It is a domino effect: If the trade winds slow, then the ocean currents also slow. This keeps the cold-water upwelling from happening. As an end result, this could threaten ecosystems because the nutrients that the ocean currents usually provide would not be delivered. The equatorial Pacific is a major fishing region, so if this were to happen, it would also be economically devastating to those countries whose livelihood is dependent on fishing resources.

PAST WARMER OCEANS

According to Karen Bice, a paleoclimatologist at the Woods Hole Oceanographic Institution, the tropical Atlantic Ocean may have once been as hot as 107°F (42°C)—much higher than today's ocean temperatures. Bice believes this event occurred millions of years ago when CO_2 levels in Earth's atmosphere reached extremely high levels. Scientists view these facts as something to take very seriously, because if it happened before, it could happen again, and because current climate models do not accommodate such a large increase in temperature (perhaps owing to a shortcoming of computer modeling that needs to be studied further and modified).

Bice cautions that warmer ocean temperatures lead to increased evaporation, which in turn could disrupt precipitation patterns. The negative effect of this scenario is that it could generate severe weather events, such as hurricanes and winter storms. Bice was able to determine the temperature of the ancient oceans through the analysis of sediment cores extracted from the seafloor. Because the sediments contained organic matter that had accumulated over tens of millions of years, she and her team were able to date them. In addition, fossilized shells contained in the sediment had their chemistry analyzed. Through this proxy data, the scientists determined that the ocean temperatures in the region got as hot as 91°–107°F (33°–42°C) approximately 84–100 million years ago.

Through the analysis of the sediments, Bice was also able to determine that the CO_2 level of the atmosphere ranged from 1,300–2,300 ppm, a significant increase compared to today's 380 ppm. One major point this study makes is how continually increasing the levels of CO_2 in the atmosphere can have a significant impact on ocean temperatures.

BREAKING THE TEMPERATURE RECORD

According to NASA scientists, 2005 was the warmest year in more than a century, followed by 1998, 2002, 2003, and 2004. Scientists also noted that in 2005, the Arctic region was unusually warm. Temperature calculations for the year are based on a multitude of temperature readings taken from field sites on land, from ships, and from satellites in space that continuously collect sea surface temperature data.

A major point, according to NASA scientists, is that in 1998 (previously the warmest year until 2005), the warming was attributed to an El Niño event, yet in 2005, there was no El Niño event involved. The atmosphere heated back up and surpassed the 1998 levels without the warming effect of an El Niño system in place. What this tells NASA scientists is that there is something else going on to cause the excessive warming experienced.

Scientists know that since the mid-1970s, global warming has warmed Earth an additional 1°F (0.6°C). Over the past century, it has warmed 1.4°F (0.8°C). According to James E. Hansen, director of NASA's GISS and a world-renowned expert on global warming, the current increase in warmth that Earth is currently experiencing is present everywhere at the same time; the largest effects are seen at the polar latitudes in the Northern Hemisphere. Over the past 50 years, the most notable seasonal warmings have occurred in Alaska, Siberia, and the Antarctic Peninsula. Furthermore, heat and pollution from urban areas cannot be blamed for this situation, because all these areas are remotely located and are not influenced by urban heat or industry.

ARCTIC MELT

According to an October 2, 2007, article in the *New York Times,* the Arctic ice cap shrank so much in the summer of 2007 that it opened two previously ice-bound historic trade routes: the Northwest Passage over Canada and the Northern Sea Route over Russia. This melt was more excessive than anything seen before, exposing 1 million cubic miles (4,168,182 km³) of open water.

This event has captured climatologists' attention because the major meltdowns had not been predicted. They determined the event was a combination of moving and melting ice. Son Nghiem, at NASA's Jet Propulsion Laboratory, is using satellite data and buoys to show that winds since 2000 have pushed huge amounts of ice out of the Arctic Basin past Greenland. Thin ice that has formed since has been easily melted or broken free by the wind and movement. What climatologists were surprised to find was that the Arctic responded faster to climate change than they initially thought.

According to NASA, what has scientists especially concerned is that they cannot find any data to indicate the Arctic has responded like this in the past; it has only been since human-induced global warming that the ice has begun to melt and shift this rapidly, adding credence to the fact that humans may have "tipped the balance" and caused irreversible climate change. There are several factors involved, including heat-trapping clouds, increased ocean warmth from clear skies, warm winds blowing off of Siberia, currents and winds pushing ice out of the Arctic Ocean, and alternating wind and pressure circulation patterns over the Arctic Ocean, called the Arctic Oscillation. They agree that more refined modeling is needed to make more definite conclusions.

As technology advances and scientists continue researching the complexities of climate change, new discoveries will continue to be made. These will help society better deal with the challenges of global warming ahead.

Conclusions and a Glance into the Future

As climatologists study the past, observe the present, and build models based on this data to predict the future, they recognize that there are still a lot of unknowns about what exactly the future will bring. Many experts have likened the situation of global warming to a real-time scientific experiment being conducted on Earth, and the actions humans take—such as burning fossil fuels—dictate how this "live experiment" will turn out. Many scientists, including the more than 2,500 represented by the IPCC, agree that using Earth and the life on it as a lab experiment is foolhardy. The potential risks are far too great no matter how much the ice sheets and glaciers melt, how great the sea-level rise by the end of the century, or how many devastating hurricanes like Katrina wreak havoc on the world's populations.

The bottom line is that the human-caused processes leading to global warming need to be dealt with now. Even if the worst effects are not felt for decades, global warming is not a problem that can be left for future generations to deal with and try to fix. In fact, no matter what the outcome of global warming is in the future, cutting back on the use of fossil fuels and other polluting activities is simply being environmentally responsible. These measures make sense on their own. While humans cannot control Earth's natural cycles and resultant climate forcings, they can control the human side of the equation. The results of prevention and responsible environmental decisions now will be felt by future generations.

This chapter looks at water resources—one of Earth's most crucial—and how global warming will impact them in the future. Next, it addresses future sea-level rise. It then presents ongoing research issues concerning future climate and global warming.

WATER RESOURCES

According to Steven Chu, director of the Lawrence Berkeley National Laboratory, water resources could be one of the first of the major, devastating impacts for life on Earth due to global warming. Chu reinforces his prediction with several cases already evident: The Yellow River in China, which is fed by glacier and snowmelt from the Himalayas, is getting noticeably lower as the glaciers melt away, for example. This presents a serious problem because a huge portion of the world's population gets its drinking water supply from these glaciers. Another example is in the United States. The snowpack in the Sierra Nevada mountains in Nevada and California are expected to decline by 30 to 70 percent by the year 2100. Chu puts this percentage loss into realistic terms, as follows:

- If snowpack declines by 20 percent, the public will be told to cut back on watering their lawns and flushing the toilet.
- If snowpack declines by 50 percent, people in California will be forced to move from drier locations to wetter locations.
- If snowpack declines more than 50 percent, the agricultural industry (fruits, vegetables, nuts, etc.) could be destroyed, and people would be forced to move out of California.

In addition, Chu says that snowfall may actually increase in some mountain areas in parts of the world as a result of global warming. Also, dry regions may become drier, while wet regions may receive more rain. He warns, however, that the extra moisture will not necessarily be advantageous. Instead, warming will prevent the extra rain and snow from being stored in the mountainous areas; most of it will run off before it can be captured, stored, and used.

Because of these dire predictions, some countries have begun to turn to the technology of seawater desalinization. In fact, General Electric has begun research into systems that can purify seawater and wastewater for human consumption. To date, however, this process is extremely expensive.

In the United States, temperatures are expected to rise everywhere, causing earlier snowmelt and more evaporation. More frequent and intense floods and droughts are possible. Because of this, some regions could benefit from increased supplies of water but could also face risks of increased flooding. Simultaneously, others could experience droughts and then floods.

Another issue that could be a problem is the current adaptation toward water resources in the United States and the distribution of dams and other water supply and irrigation infrastructure. The system

As the planet continues to warm, water resources will become scarcer and more valuable. *(Nature's Images)*

in place today is geared toward what has always been—such as the Northwest's rainy climate and the Southwest's dry climate. If climate regimes change and regions experience different climates with different water resource needs, the country may not be prepared to deal with these changes. There may not be reservoirs where they are needed to store precious water resources.

In addition, there may also be civil dissension as rivers run low and people fight for water to drink, grow crops, feed livestock, generate electricity, run industry, and complete other vital tasks. The area of the United States that has been projected to have potentially the largest negative impact is the already-dry Southwest.

SEA-LEVEL RISE

According to World Science, in a publication released March 23, 2006, sea-level rise related to global warming is producing a newly-discovered phenomenon called "glacial earthquakes." Furthermore, as asserted in a study published in *Science* magazine, Earth may be warm enough by the year 2100 for the widespread melting of the Greenland Ice Sheet and partial collapse of the Antarctic Ice Sheet.

According to Jonathan Overbeck at the University of Arizona in Tucson, based on reconstructions of past climate, sea level could rise by several yards by the end of this century. In the study, seismologists reported extensively on glacial earthquakes. These are huge glaciers that experience sudden, unexpected lurches in their forward movement. These lurches are so significant that they produce tremors on the "moment-magnitude scale" up to a magnitude of 5.1. (The moment-magnitude scale is similar to the Richter scale used in seismology to record earthquakes.) These lurches most frequently occur during July and August.

Göran Ekström, a researcher at Harvard University, says that some of these glaciers are "as large as Manhattan and as tall as the Empire State Building, and can move 33 feet (10 m) in less than a minute; which is a jolt sufficient to generate moderate seismic waves." He believes the mechanism causing this is global warming. As glaciers and the snow on them melts, water seeps downward and accumulates at the glacier's base, acting like a lubricant.

Global warming is one of the most pressing issues in the news today. Public awareness, involvement, and action are critical in order to begin the steps necessary to put an end to it. *(Nature's Images)*

Meredith Nettles, a postdoctoral researcher at Columbia University, who is also part of the research team, says these glaciers can respond to changes in climate conditions much more rapidly than previously thought. The faster they move and melt into the ocean, the faster sea level rises, and the more likely they are to slow the ocean currents that distribute warmth around the world. Greenland, which is not an area prone to seismic activity, has experienced 182 of these glacial earthquakes since 1993, with magnitudes ranging from 4.6 to 5.1. They all originated at major valleys draining the Greenland Ice Sheet. According to researchers, this indicates an increase in glacial melting, ultimately leading to sea-level rise. Evidence of glacial earthquakes has also been reported in Alaska and at the edges of Antarctica.

According to Sydney Levitus, director of NOAA's World Data Center for Oceanography, sea-level observations during the past 100 years indicate that sea level has risen at an average rate of 0.07 inch (1.7 mm) per year, most of that due to thermal expansion. Levitus and other researchers have developed models based on gathered data from 15 to 20 years ago. In their models, they have been able to separate successfully and clearly the "human signal" from Earth's "natural variability." This is enabling scientists to shift from trying to "prove" global warming is a problem to figuring out ways to fix the now-accepted problem. Some consequences of global warming are already too late to fix, even if all use of fossil fuels were stopped today. With what has already been released into the atmosphere, there will still be future warming along with its related effects. This forces not only scientists but the general public to find ways to prepare for the unstoppable effects and adjust and adapt accordingly in the future.

According to Thomas Wigley, a climate researcher at the National Center for Atmospheric Research (NCAR) in Boulder, even if the world had stopped the use of fossil fuels five years ago, global average temperature would still rise 1°F (0.6°C) by the end of the 21st century and sea levels would rise by another 4 inches (10 cm). Rising sea levels would continue well beyond 2100, even without adding meltwater from glaciers and ice sheets (because the ocean warms so slowly, it will take longer for it to respond to all the heat already added to it).

When the research team at NCAR ran their models, sea level increases continued long after temperature increases leveled out. Dr. Gerald Meehl, also at NCAR, believes that "the relentless nature of sea level rise is daunting." Meehl also says that "many people do not realize we are committed right now to a significant amount of global warming and sea level rise. The longer we wait, the more climate change we are committed to in the future."

According to Roger Pielke Jr., director of the Center for Science and Technology Policy Research at the University of Colorado at Boulder, "Prevention of global warming and subsequent sea level rise is not on the table. Adaptation needs to be given a much higher priority than it has been given. We have the scientific technological knowledge we need to improve adaptation." He recommends that science and technology apply that knowledge globally right now.

RESEARCH

The Geophysical Fluid Dynamics Laboratory at NOAA conducted modeling research centered on predicting the effects of global warming on Earth if the atmosphere's CO_2 level was doubled or quadrupled. They based their results on several century iterations of a model that combined input between both the ocean and atmosphere. In order to test and calibrate their model, they first devised and ran the past century's data through it to determine whether the model was reliable. When running old data in the model, scientists expect the model to accurately portray today's climate as its result. If it does, then the model can be trusted with current data being used to predict the future. There was a high degree of correlation between the modeled and observed global temperature curves, showing the reliability of its use in projected simulations. Results for the doubling of CO_2 predicted a rise in surface temperature of 6.2°F (3.7°C). This falls within the limits of that predicted by the IPCC, which is 2.5–7.5°F (1.5–4.5°C).

Concerning surface air temperature warming, overall warming is projected to be significant over much of the mid-latitude continental regions, such as North America and Asia. When the model was run quadrupling the CO_2 levels, scientists found that it predicted a climate similar to that which existed during the Late Cretaceous around 65 to 90 million years ago. Some of the world's hottest places under this scenario will be the United States, Europe, and Asia. During this scenario, it is also projected that the sea ice coverage over the Arctic Ocean would decrease significantly. In fact, according to NOAA, because sea ice responds quickly to the effects of greenhouse gases, it is even probable that during late summer, sea ice would be completely melted if there were four times the carbon dioxide in the atmosphere.

Concerning sea-level rise, it is expected that a thermally-driven rise in sea level will continue for centuries after atmospheric CO_2 stops increasing. With quadruple the amount of CO_2 in the atmosphere, sea level would still continue to rise 500 years into the future. The amount of sea-level rise will wreak havoc on coastal communities worldwide. Much of Florida's coasts will be inundated, putting major cities such as Miami underwater, as well as several locations along the Gulf of Mexico and up along the United States's eastern seaboard.

The thermohaline circulation (global conveyor belt) will also be negatively impacted. The model predicts the chief global heat-distributing current will decrease in intensity as the effects of global warming increase, caused mostly from ice melt in the arctic regions along with increased precipitation in high latitudes. When the CO_2 is doubled, the thermohaline weakens to less than half of its original strength, but after several centuries, returns to its original state. When the CO_2 levels are quadrupled, however, the thermohaline circulation completely shuts down in the model. Overall, the model showed that the faster the buildup of CO_2, the greater the reduction in the thermohaline circulation and the longer it took to recover, increasing the negative impacts to areas of the world such as western Europe.

Based on these results, scientists are focusing on what exactly the role is between global warming and abrupt climate change. As climatic models continue to become more refined and specialized, they will serve as tools to answer these important questions.

One of the biggest challenges in developing accurate models is the difficulty of developing models that treat the atmosphere and ocean as an integrated system. Both entities are extremely complex, and scientists still do not have all the answers. According to NOAA's Geophysical Fluid Dynamics Laboratory (GFDL), the ocean-atmosphere interface is one of the key features researchers have targeted for future development and improvement.

Current modeling has also addressed future soil moisture as it relates to dry desert regions and moist temperate zones. One of the areas researchers are currently focusing heavily on is the southwestern United States. The model projects serious decreases in soil moisture over most of the mid-latitude continental areas (such as the United States) during the summer months in both the double and quadruple CO_2 models. Under the quadruple CO_2 model, soil moisture reduction is up to 50 percent drier than today's soil moisture. If soils did reach the point where they are this much drier, scientists at GFDL say this would have a substantial impact on agricultural practices throughout much of the world's important food-producing regions. This could trigger enormous world hunger issues. This problem is so important that GFDL is currently working with scientists at the USGS to develop better predictive models.

Heat and temperature changes and their manifestation in severe weather are also being studied by the GFDL in its global warming modeling. Scientists there are currently working to refine their models to be sensitive enough to model tropical storms and hurricanes and El Niño events. The GFDL believes that by using higher resolution models, a better understanding of these weather phenomena will become available. Researchers there have begun using what they call a regional nested model—the laboratory's high-resolution Hurricane Prediction System—to explore the impact of greenhouse gas–induced climate warming on hurricane intensities.

As computer technology advances and new climatic discoveries are made, climatologists will continue to have a better picture of one of Earth's most complicated systems: weather and climate. From this knowledge we all will be able to make educated decisions about global warming that will benefit the global populations of today and tomorrow.

CHRONOLOGY

ca. 1400–1850 Little Ice Age covers the Earth with record cold, large glaciers, and snow. There is widespread disease, starvation, and death.

1800–70 The levels of CO_2 in the atmosphere are 290 ppm.

1824 Jean-Baptiste Joseph Fourier, a French mathematician and physicist, calculates that the Earth would be much colder without its protective atmosphere.

1827 Jean-Baptiste-Joseph Fourier presents his theory about the Earth's warming. At this time many believe warming is a positive thing.

1859 John Tyndall, an Irish physicist, discovers that some gases exist in the atmosphere that block infrared radiation. He presents the concept that changes in the concentration of atmospheric gases could cause the climate to change.

1894 Beginning of the industrial pollution of the environment.

1913–14 Svante Arrhenius discovers the greenhouse effect and predicts that the Earth's atmosphere will continue to warm. He predicts that the atmosphere will not reach dangerous levels for thousands of years, so his theory is not received with any urgency.

1920–25 Texas and the Persian Gulf bring productive oil wells into operation, which begins the world's dependency on a relatively inexpensive form of energy.

1934 The worst dust storm of the dust bowl occurs in the United States on what historians would later call Black Sunday. Dust storms are a product of drought and soil erosion.

1945 The U.S. Office of Naval Research begins supporting many fields of science, including those that deal with climate change issues.

1949–50   Guy S. Callendar, a British steam engineer and inventor, propounds the theory that the greenhouse effect is linked to human actions and will cause problems. No one takes him too seriously, but scientists do begin to develop new ways to measure climate.

1950–70   Technological developments enable increased awareness about global warming and the enhanced greenhouse effect. Studies confirm a steadily rising CO_2 level. The public begins to notice and becomes concerned with air pollution issues.

1958   U.S. scientist Charles David Keeling of the Scripps Institution of Oceanography detects a yearly rise in atmospheric CO_2. He begins collecting continuous CO_2 readings at an observatory on Mauna Loa, Hawaii. The results became known as the famous Keeling Curve.

1963   Studies show that water vapor plays a significant part in making the climate sensitive to changes in CO_2 levels.

1968   Studies reveal the potential collapse of the Antarctic ice sheet, which would raise sea levels to dangerous heights, causing damage to places worldwide.

1972   Studies with ice cores reveal large climate shifts in the past.

1974   Significant drought and other unusual weather phenomenon over the past two years cause increased concern about climate change not only among scientists but with the public as a whole.

1976   Deforestation and other impacts on the ecosystem start to receive attention as major issues in the future of the world's climate.

1977   The scientific community begins focusing on global warming as a serious threat needing to be addressed within the next century.

1979   The World Climate Research Programme is launched to coordinate international research on global warming and climate change.

1982 Greenland ice cores show significant temperature oscillations over the past century.

1983 The greenhouse effect and related issues get pushed into the political arena through reports from the U.S. National Academy of Sciences and the Environmental Protection Agency.

1984–90 The media begins to make global warming and its enhanced greenhouse effect a common topic among Americans. Many critics emerge.

1987 An ice core from Antarctica analyzed by French and Russian scientists reveals an extremely close correlation between CO_2 and temperature going back more than 100,000 years.

1988 The United Nations set up a scientific authority to review the evidence on global warming. It is called the Inter-governmental Panel on Climate Change (IPCC) and consists of 2,500 scientists from countries around the world.

1989 The first IPCC report says that levels of human-made greenhouse gases are steadily increasing in the atmosphere and predicts that they will cause global warming.

1990 An appeal signed by 49 Nobel prizewinners and 700 members of the National Academy of Sciences states, "There is broad agreement within the scientific community that amplification of the Earth's natural greenhouse effect by the buildup of various gases introduced by human activity has the potential to produce dramatic changes in climate . . . Only by taking action now can we insure that future generations will not be put at risk."

1992 The United Nations Conference on Environment and Development (UNCED), known informally as the Earth Summit, begins on June 3 in Rio de Janeiro, Brazil. It results in the United Nations Framework Convention on Climate Change, Agenda 21, the Rio Declaration on Environment and Development Statement of Forest Principles, and the United Nations Convention on Biological Diversity.

1993 Greenland ice cores suggest that significant climate change can occur within one decade.

1995 The second IPCC report is issued and concludes there is a human-caused component to the greenhouse effect warming. The consensus is that serious warming is likely in the coming century. Reports on the breaking up of Antarctic ice sheets and other signs of warming in the polar regions are now beginning to catch the public's attention.

1997 The third conference of the parties to the Framework Convention on Climate Change is held in Kyoto, Japan. Adopted on December 11, a document called the Kyoto Protocol commits its signatories to reduce emissions of greenhouse gases.

2000 Climatologists label the 1990s the hottest decade on record.

2001 The IPPC's third report states that the evidence for anthropogenic global warming is incontrovertible, but that its effects on climate are still difficult to pin down. President Bush declares scientific uncertainty too great to justify Kyoto Protocol's targets.

The United States Global Change Research Program releases the findings of the National Assessment of the Potential Consequences of Climate Variability and Change. The assessment finds that temperatures in the United States will rise by 5 to 9°F (3–5°C) over the next century and predicts increases in both very wet (flooding) and very dry (drought) conditions. Many ecosystems are vulnerable to climate change. Water supply for human consumption and irrigation is at risk due to increased probability of drought, reduced snow pack, and increased risk of flooding. Sea-level rise and storm surges will most likely damage coastal infrastructure.

2002 Second hottest year on record.

Heavy rains cause disastrous flooding in Central Europe leading to more than 100 deaths and more than $30 billion in damage. Extreme drought in many parts of the world (Africa, India,

Australia, and the United States) results in thousands of deaths and significant crop damage. President Bush calls for 10 more years of research on climate change to clear up remaining uncertainties and proposes only voluntary measures to mitigate climate change until 2012.

2003 U.S. senators John McCain and Joseph Lieberman introduce a bipartisan bill to reduce emissions of greenhouse gases nation-wide via a greenhouse gas emission cap and trade program.

Scientific observations raise concern that the collapse of ice sheets in Antarctica and Greenland can raise sea levels faster than previously thought.

A deadly summer heat wave in Europe convinces many in Europe of the urgency of controlling global warming but does not equally capture the attention of those living in the United States.

International Energy Agency (IEA) identifies China as the world's second largest carbon emitter because of their increased use of fossil fuels.

The level of CO_2 in the atmosphere reaches 382 ppm.

2004 Books and movies feature global warming.

2005 Kyoto Protocol takes effect on February 16. In addition, global warming is a topic at the G8 summit in Gleneagles, Scotland, where country leaders in attendance recognize climate change as a serious, long-term challenge.

Hurricane Katrina forces the U.S. public to face the issue of global warming.

2006 Former U.S. vice president Al Gore's *An Inconvenient Truth* draws attention to global warming in the United States.

Sir Nicholas Stern, former World Bank economist, reports that global warming will cost up to 20 percent of worldwide gross domestic product if nothing is done about it now.

2007 IPCC's fourth assessment report says glacial shrinkage, ice loss, and permafrost retreat are all signs that climate change is underway now. They predict a higher risk of drought, floods, and more powerful storms during the next 100 years. As a result, hunger, homelessness, and disease will increase. The atmosphere may warm 1.8 to 4.0°C and sea levels may rise 7 to 23 inches (18 to 59 cm) by the year 2100.

Al Gore and the IPCC share the Nobel Peace Prize for their efforts to bring the critical issues of global warming to the world's attention.

2008   The price of oil reached and surpassed $100 per barrel, leaving some countries paying more than $10 per gallon.

Energy Star appliance sales have nearly doubled. Energy Star is a U.S. government-backed program helping businesses and individuals protect the environment through superior energy efficiency.

U.S. wind energy capacity reaches 10,000 megawatts, which is enough to power 2.5 million homes.

2009   President Obama takes office and vows to address the issue of global warming and climate change by allowing individual states to move forward in controlling greenhouse gas emissions. As a result, American automakers can prepare for the future and build cars of tomorrow and reduce the country's dependence on foreign oil. Perhaps these measures will help restore national security and the health of the planet, and the U.S. government will no longer ignore the scientific facts.

The year 2009 will be a crucial year in the effort to address climate change. The meeting on December 7–18 in Copenhagen, Denmark, of the UN Climate Change Conference promises to shape an effective response to climate change. The snapping of an ice bridge in April 2009 linking the Wilkins Ice Shelf (the size of Jamaica) to Antarctic islands could cause the ice shelf to break away, the latest indication that there is no time to lose in addressing global warming.

LINKS TO GLOBAL WARMING SITES

GLOBAL WARMING

Climate Ark. Available online. URL: www.climateark.org. Accessed June 17, 2009. Promotes public policy to address global climate change through reduction of carbon and other emissions, energy conservation, alternative energy sources, and ending deforestation.

Climate Solutions. Available online. URL: www.climatesolutions.org. Accessed June 17, 2009. Practical solutions to global warming.

Environmental Defense Fund. Available online. URL: www.environmentaldefense.org. Accessed June 17, 2009. An organization started by a handful of environmental scientists in 1967 that provides quality information and helpful resources on understanding global warming and other crucial environmental issues.

Environmental Protection Agency. Available online. URL: www.epa.gov. Accessed June 17, 2009. Provides information about EPA's efforts and programs to protect the environment; it offers a wide array of information on global warming.

European Environment Agency. Sponsored by the European Environment Agency in Copenhagen, Denmark. Available online. URL: www.eea.europa.eu/themes/climate. Accessed June 17, 2009. Posts their reports on topics such as air quality, ozone depletion, and climate change.

Global Warming: Focus on the Future. Available online. URL: www.enviroweb.org. Accessed June 17, 2009. Offers statistics and photography of global warming topics.

HotEarth.Net. Available online. URL: www.net.org/warming. Accessed June 17, 2009. Features informational articles on the causes of global warming, its harmful effects, and solutions that could stop it.

Intergovernmental Panel on Climate Change (IPCC). Available online. URL: http://www.ipcc.ch/. Accessed June 17, 2009. Offers current information on the science of global warming and recommendations on practical solutions and policy management.

NASA's Goddard Institute for Space Studies. Available online. URL: www.giss.nasa.gov. Accessed June 17, 2009. This Web site provides a large database of information, research, and other resources.

NOAA's National Climatic Data Center. Available online. URL: www.ncdc.noaa.gov. Accessed June 17, 2009. Offers a multitude of resources and information on climate, climate change, global warming.

Ozone Action. Available online. URL: www.semcog.org/Ozone Action_kids.aspx. Accessed June 17, 2009. Provides information on air quality by focusing on ozone, the atmosphere, environmental issues, and related health issues.

Scientific American. Available online. URL: www.sciam.com. Accessed June 17, 2009. An online magazine that often presents articles concerning climate change and global warming.

Tyndall Centre at University of East Anglia. Available online. URL: http://www.tyndall.ac.uk. Accessed June 17, 2009. Offers information on climate change and is considered one of the leaders in UK research on global warming.

Union of Concerned Scientists. Available online. URL: www.ucsusa.org. Accessed June 17, 2009. This Web site offers a quality resource sections on global warming and ozone depletion.

United Nations Framework Convention on Climate Change (UNFCCC). Sponsored by the United Nations Framework Convention on Climate Change. Available online. URL: http://unfccc.int/2860.php. Accessed June 17, 2009. A spectrum on climate change information and policy.

U.S. Global Change Research Program. Available online. URL: www.usgcrp.gov. Accessed June 17, 2009. Provides information on the current research activities of national and international science programs that focus on global monitoring of climate and ecosystem issues.

World Wildlife Foundation Climate Change Campaign. Available online. URL: www.worldwildlife.org/climate/. Accessed June 17, 2009. Contains information on what various countries are doing and not doing to deal with global warming.

GREENHOUSE GAS EMISSIONS

Energy Information Administration. Available online. URL: www.eia. doe.gov/environment.html. Accessed June 17, 2009. Lists official environmental energy-related emissions data and environmental analyses from the U.S. government. Contains carbon dioxide, methane, and nitrous oxide emissions data and other greenhouse gas reports.

World Resources Institute—Climate, Energy & Transport. Available online. URL: www.wri.org/climate/publications.cfm. Accessed June 17, 2009. This Web site offers a collection of reports on global technology deployment to stabilize emissions, agriculture, and greenhouse gas mitigation, climate science discoveries, and renewable energy.

GLOSSARY

adaptation   An adjustment in natural or human systems to a new or changing environment. Adaptation to climate change refers to adjustments in natural or human systems in response to actual or expected climatic changes.

aerosols   Tiny bits of liquid or solid matter suspended in air. They come from natural sources such as erupting volcanoes and from waste gases emitted from automobiles, factories, and power plants. By reflecting sunlight, aerosols cool the climate and offset some of the warming caused by greenhouse gases.

albedo   The relative reflectivity of a surface. A surface with high albedo reflects most of the light that shines on it and absorbs very little energy; a surface with a low albedo absorbs most of the light energy that shines on it and reflects very little.

altimeter   A sensitive aneroid barometer that is graduated and calibrated, used chiefly for finding distance above sea level, terrain, or some other reference point.

anemometer   Any instrument for measuring the speed of wind.

anthropogenic   Made by people or resulting from human activities. This term is usually used in the context of emissions that are produced as a result of human activities.

atmosphere   The thin layer of gases that surround the Earth and allow living organisms to breathe. It reaches 400 miles (644 km) above the surface, but 80 percent is concentrated in the troposphere—the lower seven miles (11 km) above the Earth's surface.

biodiversity   Different plant and animal species.

carbon   A naturally abundant nonmetallic element that occurs in many inorganic and in all organic compounds.

carbon dioxide   A colorless, odorless gas that passes out of the lungs during respiration. It is the primary greenhouse gas and causes the greatest amount of global warming.

189

carbon sink   An area where large quantities of carbon are built up in the wood of trees, in calcium carbonate rocks, in animal species, in the ocean, or any other place where carbon is stored. These places act as reservoirs, keeping carbon out of the atmosphere.

climate   The usual pattern of weather that is averaged over a long period of time.

climate model   A quantitative way of representing the interactions of the atmosphere, oceans, land surface, and ice. Models can range from relatively simple to extremely complicated.

climatologist   A scientist who studies the climate.

concentration   The amount of a component in a given area or volume. In global warming, it is a measurement of how much of a particular gas is in the atmosphere compared to all of the gases in the atmosphere.

condense   The process that changes a gas into a liquid.

cyclone   A large-scale, atmospheric wind and pressure system characterized by low pressure at the center and circular wind motion, counterclockwise in the Northern Hemisphere, clockwise in the Southern Hemisphere.

deforestation   The large-scale cutting of trees from a forested area, often leaving bare areas susceptible to erosion.

ecosystem   A community of interacting organisms and their physical environment.

emissions   The release of a substance (usually a gas when referring to the subject of climate change) into the atmosphere.

evaporation   The process by which a liquid, such as water, is changed to a gas.

feedback   A change caused by a process that, in turn, may influence that process. Some changes caused by global warming may hasten the process of warming (positive feedback); some may slow warming (negative feedback).

forcings   Mechanisms that disrupt the global energy balance between incoming energy from the Sun and outgoing heat from the Earth.

By altering the global energy balance, such mechanisms force the climate to change. Today, anthropogenic greenhouse gases added to the atmosphere are forcing climate to change.

fossil fuel   An energy source made from coal, oil, or natural gas. The burning of fossil fuels is one of the chief causes of global warming.

glacier   A mass of ice formed by the buildup of snow over hundreds and thousands of years.

global dimming   A reduction in the amount of the Sun's electromagnetic energy reaching the Earth's surface due to its blockage by particulate matter, clouds, and other opaque materials in the atmosphere.

global warming   An increase in the temperature of the Earth's atmosphere, caused by the buildup of greenhouse gases. This is also referred to as the enhanced greenhouse effect caused by humans.

great ocean conveyor belt   A global current system in the ocean that transports heat from one area to another.

greenhouse effect   The natural trapping of heat energy by gases present in the atmosphere, such as CO_2, methane, and water vapor. The trapped heat is then emitted as heat back to the Earth.

greenhouse gas   A gas that traps heat in the atmosphere and keeps the Earth warm enough to allow life to exist.

Gulf Stream   A warm current that flows from the Gulf of Mexico across the Atlantic Ocean to northern Europe. It is largely responsible for Europe's milder climate.

hydrologic cycle   The natural sequence through which water passes into the atmosphere as water vapor, precipitates to earth in liquid or solid form, and ultimately returns to the atmosphere through evaporation.

Industrial Revolution   The period during which industry developed rapidly as a result of advances in technology. This took place in Britain during the late 18th and early 19th centuries.

infrared   The invisible heat radiation that is emitted by the Sun and by virtually every warm substance or object on Earth.

interdecadal   A time period that occurs within a decade's time.

IPCC   Intergovernmental Panel on Climate Change. This is an organization consisting of 2,500 scientists that assesses information in the scientific and technical literature related to the issue of climate change. The IPCC was established jointly by the United Nations Environment Programme and the World Meteorological Organization in 1988.

jet stream   A strong ribbon of horizontal wind that is found about 6 to 10 miles (10–16 km) above the ground in the area between the troposphere, the lower layer of the atmosphere, and the stratosphere above it.

Keeling Curve   A famous curve showing increasing CO_2 concentrations in the atmosphere which was set up by Dr. Charles David Keeling of Scripps Institution of Oceanography at Mauna Loa in Hawaii, it illustrate the steady rise in CO_2 concentrations since 1958.

Maunder minimum   The period of reduced solar activity lasting through the 1600s and 1700s.

methane   A colorless, odorless, flammable gas that is the major ingredient of natural gas. Methane is produced wherever decay occurs and little or no oxygen is present.

multidecadal   A time span over several decades, such as 50 to 80 years.

nitrous oxide   A heat-absorbing gas in the Earth's atmosphere. Nitrous oxide is emitted from nitrogen-based fertilizers.

parts per million (ppm)   The number of parts of a chemical found in one million parts of a particular gas, liquid, or solid.

permafrost   Permanently frozen ground in the Arctic. As global warming increases, this ground is melting.

photosynthesis   The process by which plants make food using light energy, carbon dioxide, and water.

protocol   The terms of a treaty that have been agreed to and signed by all parties.

proxies   Methods of determining values such as temperatures and rainfall by using substitutes, which give indirect measurements. Tree rings serve as proxies for determining rainfall abundance.

radiation   The particles or waves of energy.

renewable   Something that can be replaced or regrown, such as trees, or a source of energy that never runs out, such as solar energy, wind energy, or geothermal energy.

resources   The raw materials from the Earth that are used by humans to make useful things.

rotation   The movement or path of the Earth, turning on its axis.

satellite   Any small object that orbits a larger one. Artificial satellites carry instruments for scientific study and communication. Imagery taken from satellites is used to monitor aspects of global warming such as glacier retreat, ice cap melting, desertification, erosion, hurricane damage, and flooding. Sea surface temperatures and measurements are also obtained from man-made satellites in orbit around the Earth.

simulation   A computer model of a process that is based on actual facts. The model attempts to mimic, or replicate, actual physical processes.

stratosphere   The layer of the atmosphere just above the troposphere. It extends 7.5 miles (12 km) to an average of 31 miles (50 km).

subsidence   To sink to a lower level, such as the ground.

temperate   An area that has a mild climate and different seasons.

thermal   Something that relates to heat.

trade winds   Winds that blow steadily from east to west and toward the equator. The trade winds are caused by hot air rising at the equator, with cool air moving in to take its place from the north and from the south. The winds are deflected westward because of the Earth's west-to-east rotation.

tropical   A region that is hot and often wet (humid). These areas are located around the Earth's equator.

tropical storm   A cyclonic storm having winds ranging from approximately 30 to 75 miles (48–121 kilometers) per hour.

tundra   A vast treeless plain in the Arctic with a marshy surface covering a permafrost layer.

upwelling   The process by which warm, less-dense surface water is drawn away from along a shore by offshore currents and replaced by cold, denser water brought up from the subsurface.

weather   The conditions of the atmosphere at a particular time and place. Weather includes such measurements as temperature, precipitation, air pressure, and wind speed and direction.

weathering   The progression of breaking down rocks and natural materials on the Earth's surface through physical and chemical processes.

westerlies   A semipermanent belt of westerly winds that prevails at latitudes lying between the tropical and polar regions of the Earth.

FURTHER RESOURCES

BOOKS

Christianson, Gale. *Greenhouse: The 200-Year Story of Global Warming.* New York: Walker, 1999. This book looks at the enhanced greenhouse effect worldwide since the Industrial Revolution and outlines the consequences to the environment.

Friedman, Katherine. *What If the Polar Ice Caps Melted?* Danbury, Conn.: Children's Press, 2002. This book focuses on environmental problems related to the Earth's atmosphere, including global warming, changing weather patterns, and their effects on ecosystems.

Gelbspan, Ross. *The Heat Is On: The High Stakes Battle Over Earth's Threatened Climate.* Reading, Mass.: Addison Wesley, 1997. This work offers a look at the controversy environmentalists often face when they deal with fossil fuel companies.

Harrison, Patrick, Gail "Bunny" McLeod. *Who Says Kids Can't Fight Global Warming.* Chattanooga, Tenn.: Pat's Top Products, 2007. This book offers real solutions that everybody can do to help solve the world's biggest air pollution problems.

Houghton, John. *Global Warming: The Complete Briefing.* New York: Cambridge University Press, 2004. This book outlines the scientific basis of global warming and describes the impacts that climate change will have on society. It also looks at solutions to the problem.

Langholz, Jeffrey. *You Can Prevent Global Warming (and Save Money!): 51 Easy Ways.* Riverside, N.J.: Andrews McMeel Publishing, 2003. This book aims to convert public concern over global warming into positive action by providing simple, everyday practices that can easily be done to minimize it, as well as save the person money.

McKibben, Bill. *Fight Global Warming Now: The Handbook for Taking Action in Your Community.* New York: Holt Paperbacks, 2007. This book provides the facts of what must change to save the climate. It also shows how people can act proactively in their community to make a difference.

Pringle, Laurence. *Global Warming: The Threat of Earth's Changing Climate.* New York: SeaStar Publishing, 2001. This book provides information on the carbon cycle, rising sea levels, El Niño, aerosols, smog, flooding, and other issues related to global warming.

Thornhill, Jan. *This Is My Planet: The Kids Guide to Global Warming.* Toronto, Canada: Maple Tree Press, 2007. This book offers students the tools they need to become ecologically oriented by taking a comprehensive look at climate change in polar, ocean, and land-based ecosystems.

Weart, Spencer R. *The Discovery of Global Warming.* Cambridge, Mass.: Harvard University Press, 2004. This book traces the history of the global warming concept through a long process of incremental research rather than a dramatic revelation.

PRINT AND ONLINE ARTICLES

Alley, Richard B. "Abrupt Climate Change." *Scientific American* (November 2004), 62–69. This article introduces the concept of the shutting down of the thermohaline circulation, causing an ice age.

Amos, Jonathan. "Arctic Summers Ice-Free by 2013." BBC News. December 12, 2007. Available online. URL: http://news.bbc.co.uk/2/hi/science/nature/7139797.stm. Accessed June 17, 2009. Discusses melting of the ice and the Northwest Passage.

Boswell, Randy. "Northwest Passage in Unprecedented Ice Melt, Experts Report." CanWest News Service. August 28, 2007. Available online. URL: http://www.canada.com/cityguides/fortstjohn/story.html?id=3bf042a8-3bad-4728-90f3-dd58 cda33244&k=44943. Accessed June 17, 2009. Discusses current modeling techniques and causes of rapid ice melting.

Boyd, Robert S. "Glaciers Melting Worldwide, Study Finds." National Geographic News. August 21, 2002. Available online. URL: http://news.nationalgeographic.com/news/2002/08/0821_020821_

wireglaciers.html. Accessed June 17, 2009. This looks at a variety of glaciers worldwide that are melting.

Britt, Robert Roy. "Scientists Put Melting Mystery on Ice." *MSNBC.* June 30, 2005. Available online. URL: http://www.msnbc.msn.com/ id/8421342/. Accessed June 17, 2009. Discusses the rapid melting in the arctic regions.

Culotta, Elizabeth. "Will Plants Profit from High CO_2?" *Science* (May 5, 1995), 654–656. This article explores the possible effects of various carbon dioxide levels on vegetation as a result of global warming and whether it will experience enhanced growth.

D'Agnese, Joseph. "Why Has Our Weather Gone Wild?" *Discover* (June 2000), 72–78. This article focuses on the recent global changes in weather, such as shifting seasons, severe storms, droughts, heat waves, and other weather-related events, and their connection to global warming.

Eilperin, Juliet. "Severe Hurricanes Increasing, Study Finds." *Washington Post,* September 16, 2005. Available online. URL: http:// www.washingtonpost.com/wp-dyn/content/article/2005/09/15/ AR2005091502234.html. Accessed June 17, 2009. This article explores the connection between the destructiveness of hurricanes and climate change.

Emanuel, K. "Increasing Destructiveness of Tropical Cyclones Over the Past 30 Years." *Nature* (July 31, 2005). Available online. URL: http://www.nature.com/nature/journal/v436/n7051/full/ nature03906.html. Accessed June 17, 2009. This study presents evidence as to whether hurricanes are stronger because of the influence of global warming.

Gordon, Carolyn. "Tracking Glacial Activity in Norway with Photogrammetry Software." *Imaging Notes* 22, no. 1 (Spring, 2007): 24–29. Presents a method, using aerial photography and 3-D modeling techniques, to measure the mass melting of glaciers so as to calculate their contribution to sea-level rise.

Hanley, Charles J. "Glaciers Are Vanishing Around the World." *USA Today,* January 27, 2005. Available online. URL: http://www.usa today.com/weather/resources/climate/2005-01-27-warming-1-30-

glaciers_x.htm. Accessed June 17, 2009. Examines the plight of glaciers worldwide under the effects of rising temperatures.

Henderson, Mark. "Global Warming Linked to Increase of Hurricanes." *Times,* September 16, 2005. Available online. URL: http://
www.timesonline.co.uk/tol/news/world/article567156.ece. Accessed
June 17, 2009. This article discusses the increasing frequency of category 4 and 5 storms.

Hoffman, Paul F., and Daniel P. Schrag. "Snowball Earth." *Scientific
American* (January 2000), 68–75. This article presents evidence that
global climate change put the Earth in the grip of a complete frozen
state eons ago.

Hogan, Jenny. "Antarctic Ice Sheet Is an 'Awakened Giant.'" *New Scientist* (February 2, 2005). Available online. URL: http://www.new
scientist.com/article/dn6962-antarctic-ice-sheet-is-an-awakened-
giant.htm. Accessed March 14, 2009. This article discusses the melting of Antarctica and the global consequences it would have.

Karl, Thomas, and Kevin Trenberth. "The Human Impact on Climate." *Scientific American* (December 1999), 100–105. This article
focuses on the disruptions people cause in the natural environment
and why scientists must begin to monitor and quantify the disruptions now in order to save the future.

Knutson, T. R., and R. E. Tuleya. "Impact of CO_2-induced Warming
on Simulated Hurricane Intensity and Precipitation: Sensitivity to
the Choice of Climate Model and Convective Parameterization."
Journal of Climate 20, no. 13 (2004): 2,961. This work explores climate models and extreme weather.

Renfrow, Stephanie. "Arctic Sea Ice on the Wane." Earth System Science Data and Services. October 31, 2006. Available online. URL:
http://nasadaacs.eos.nasa.gov/articles/2006/2006_seaice.html.
Accessed June 17, 2009. This discusses recent sea ice melting and
the opening of the Northwest Passage.

Revkin, Andrew D. "Arctic Melt Unnerves the Experts." *New York
Times,* October 2, 2007. Available online. URL: http://yaleglobal.
yale.edu/display.article?id=9812. Accessed June 17, 2009. This arti-

cle discusses rapidly rising temperatures and the consequences of passing "tipping points" at which sea-level rise will become destructive to urban areas and ecosystems.

———. "In Greenland, Ice and Instability." *New York Times,* January 8, 2008. Available online. URL: http://www.nytimes.com/2008/01/08/science/earth/08gree.html. Accessed June 17, 2009. This reviews the short- and long-term outlooks for Greenland and its ice resources.

———. "Melting Ice-Rising Seas? Easy. How Fast? Hard." *New York Times.* January 8, 2008. Available online. URL: http://dotearth.blogs.nytimes.com/2008/01/08/melting-ice-rising-seas-easy-how-fast-hard. Accessed June 17, 2009. This article discusses the consequences of rising sea levels, their disastrous effects, and the frustrations of trying to get others to take global warming seriously.

Roach, John. "Is Global Warming Making Hurricanes Worse?" National Geographic News. August 4, 2005. Available online. URL: http://news.nationalgeographic.com/news/2005/08/0804_050804_hurricanewarming.html. Accessed June 17, 2009. This work explores global warming's possible influence on extreme weather.

Schirber, Michael. "Will Global Warming Make Hurricanes Stronger?" MSNBC. June 16, 2005. Available online. URL: http://www.msnbc.msn.com/id/8245668/. Accessed June 17, 2009. Provides a close look at the correlations between hurricanes and global warming.

Time, eds. "51 Things We Can Do to Save the Environment." *Time* (April 9, 2007). Available online. URL: http://www.time.com/time/specials/2007/environment/. Accessed June 17, 2009. The main article of this special edition suggests 51 ways to save the environment and curb global warming.

Trenberth, Kevin E. "Warmer Oceans, Stronger Hurricanes." *Scientific American* (July 2007). Available online. URL: http://www.sciam.com/article.cfm?id=warmer-oceans-stronger-hurricanes. Accessed June 17, 2009. This looks at how global warming is enhancing cyclones' damaging winds and flooding rains.

U.S. Geological Survey. "Fact Sheet 2005-3055, Coastal-Change and Glaciological Maps of Antarctica." May 2005, revised 2007.

Available online. URL: http://pubs.usgs.gov/fs/2005/3055/index. html. Accessed June 17, 2009. Shows a graphic representation of the changes in ice cover.

Webster, P. J., et al. "Changes in Tropical Cyclone Number, Duration, and Intensity in a Warming Environment." *Science* (2005). Available online. URL: http://www.sciencemag.org/cgi/content/full/309/5742/1844. Accessed June 17, 2009. This looks at the possibility that global warming directly affects hurricanes and their intensities.

Williams, Jack. "Answers to Sea Level Rise Locked in Ice." *USA Today,* January 19, 1999. Available online. URL: http://www.usatoday. com/weather/resources/coldscience/aicesheet.htm. Accessed June 17, 2009. This article looks at paleodating techniques.

INDEX